A LETTER TO LIBERALS

A LETTER TO LIBERALS

from
ROBERT F. KENNEDY JR.

Censorship and COVID: An Attack on Science and American Ideals

Skyhorse Publishing

Children's
Health Defense

Skyhorse Publishing books may be purchased in bulk at special discounts for sales promotion, corporate gifts, fund-raising, or educational purposes. Special editions can also be created to specifications. For details, contact the Special Sales Department, Skyhorse Publishing, 307 West 36th Street, 11th Floor, New York, NY 10018 or info@skyhorsepublishing.com.

Skyhorse® and Skyhorse Publishing® are registered trademarks of Skyhorse Publishing, Inc.®, a Delaware corporation.

Visit our website at www.skyhorsepublishing.com.

10 9 8 7 6 5

Library of Congress Cataloging-in-Publication Data is available on file.

Cover design by Brian Peterson

ISBN: 978-1-5107-7558-9
eBook ISBN: 978-1-5107-7559-6

Printed in the United States of America

Contents

AUTHOR'S NOTE

[A]ttacks on me, quite frankly, are attacks on science. . . . So if you are trying to get at me as a public health official and scientist, you're really attacking not only Dr. Anthony Fauci, you are attacking science. . . . You have to be asleep not to see that.
—NIAID director Anthony Fauci, *Meet the Press*, June 9, 2021

It is troubling enough that our country's leading public health technocrat and the fiat leader of the National Institute of Health (NIH)—the world's principal funder of scientific research—would make such a narcissistic and scientifically absurd statement. The more serious concern is that the majority of my political party—the Democrats—and the mainstream media generally accept Dr. Fauci's assertion as gospel. Journalists—even science journalists—act as if they believe that any pronouncement by Dr. Anthony Fauci (or FDA, CDC, or WHO) should mark the end of scientific inquiry. It is my hope that this short book will remind all Americans that blind faith in authority is a feature of religion and autocracy, but not of science nor democracy.

In what was arguably one of the most important speeches in American history, President Dwight D. Eisenhower warned our citizenry precisely against this kind of misplaced faith in federal scientific bureaucrats:

> *The potential for the disastrous rise of misplaced power exists and will persist. . . . We must never let the weight of this combination endanger our liberties or democratic processes. In this revolution, research has become central; it also becomes more formalized, complex, and costly. A steadily increasing share is conducted for, by, or at the direction of, the Federal government. . . . The prospect of domination of the nation's scholars by Federal employment, project allocations, and the power of*

money is ever present and is gravely to be regarded.
We must . . . be alert to the . . . danger that
public policy could itself become the captive
of a scientific-technological elite.

This essay emerged from a congenial and ongoing conversation, during the COVID pandemic, between myself and my longtime friend and former law partner, John Morgan, a lifelong champion of the Democratic Party and liberal values.

I invited John—who reveres Anthony Fauci and accepts the scientific validity of the government's COVID countermeasures—to reengage his critical thinking skills and to accept my challenge to science-based debate, which he did. I hope this little book will encourage other liberal Democrats to do the same.

                              Robert F. Kennedy Jr.

A Challenge to Debate

My dear fellow Liberal,

Just before his death in 1642, Galileo complained that the authors of his 1615 censure were not just the clergy—understandably fearful that heliocentrism would subvert Church cosmologies—but, oddly, his fellow scientists, who universally refused to look through his telescope.

I am an FDR/Kennedy liberal, but my choice to openly question government policies for managing the pandemic—under both Presidents Biden and Trump—has made me pariah, primarily in liberal circles. Many traditional liberals—reacting to the orchestrated fear and propaganda—have embraced "Lockdown Liberalism," an ideology that departs dramatically from the tenets of traditional liberalism. Like Galileo's colleagues, so many of today's "Lockdown Liberals" refuse to read or debate the science that they *believe* supports the government's COVID countermeasures. Instead, they place their faith in the official orthodoxies of famously corrupt pharmaceutical companies and their notoriously captive federal agencies and expect others to do the same. This blind obedience is itself a kind of novel virus that now infects the entire upper deck of the Democratic Party. The core of this ideology is a cult-like fealty to COVID-19 countermeasures that are, in fact, scientifically indefensible. By necessity then, the acolytes of this theology must be ferociously hostile toward debate that might expose errors in government dogma and must, like the Roman Inquisition that extracted Galileo's recantation under threat of burning at the stake, mercilessly suppress every utterance of heresy or dissent. Moreover, Lockdown Liberalism's enthusiastic embrace of censorship—once anathema to liberals—has expanded into a repudiation of almost all the precepts of traditional FDR/Kennedy liberalism.

This letter is a challenge to my fellow liberals to reexamine the scientific assertions upon which rest the oppressive policies that have savaged the presumptions of classical liberalism and the United States Constitution. It is past time that our nation had an open conversation about the strategies supposedly enacted for ending the pandemic, and the best measures for avoiding future crises.

An Incongruous Liberal Allergy to Debate

The word "liberal" derives from the Latin *liber*, which the Etymology Dictionary renders as "freedom from restraint in speech or action" and "freedom from bigotry." Conventional FDR/JFK liberalism prided itself on its open-minded tolerance of contrary opinion, its implacable protectiveness of the right to dissent, its embrace of new ideas, and its fearless love for contention and disputation. Democrats were once the party of intellectual curiosity, critical thinking, and faith in scientific and liberal empiricism. Liberalism's foundational assumption, after all, is that freedom of speech and expression are essential to a functioning democracy; the free flow of information yields governing policies that have been annealed in the cauldron of fierce, open debate before triumphing on the battle-field of ideas.

We Democrats once took pride in ourselves as the party that understood how to read science critically. We confronted—and mercilessly deconstructed—the fatally flawed faux-science contrived by the carbon industry's PhD biostitutes to support climate change denialism. We also exercised healthy skepticism toward the corrupt drug companies that brought us the opioid crisis and that have paid $86 billion in criminal and civil penalties for a wide assortment of frauds and other crimes since 2000.[1] We were disgusted by the phenom-enon of "agency capture" and felt a deep revulsion for Pharma's pervasive control of Congress, the media, and the scientific journals. How is it, then, that today's Democrats become angry at the mere suggestion that the prevailing COVID drug and vaccine narrative may be heavily manipulated through orchestrated propa-ganda by a Pharma cartel with billions at stake in pro-moting COVID countermeasures?

According to an August 18, 2021, Pew Research Center Survey, 65 percent of Democrats currently support government censorship of unauthorized opinions.[2] That astonishing result suggests that Democrats have lost their faith not only in their party traditions, but also in democracy. The majority of Democrats appear to believe that the *Demos*—the people—can no longer be trusted to govern themselves and that it is, therefore, permissible for elites to manipulate the public with propaganda, and even to censor information that might infect the population with dangerous thoughts.

Liberals have long agreed that censorship of dissent is the emblem of totalitarian systems. The new strategy of silencing government critics like myself is therefore repugnant to liberalism's foundational values and is clearly offensive to the American Constitution's guarantee of free speech.

Like Galileo's colleagues, the "Lockdown Left" has abandoned the discipline of evidence-based medicine. Instead of scientific citation, they rely on appeals to often undeserving authorities who have manufactured "scientific consensus" by cherry-picking data to support a predetermined policy. Sanctimonious bromides to "follow the science," "trust the experts," most often mean blind dogmatic trust in the official—and often whimsical—pronouncements of amoral pharmaceutical companies and their venal government vassals at captive agencies like CDC, FDA, NIH, and WHO.

Unable to defend the scientific underpinnings of their ideology in debate, liberals rely on book bans and an arsenal of coercive muzzling strategies including deplatforming, delicensing, doxxing, gaslighting, defunding, retracting, marginalizing, and vilifying scientists, physicians, journalists, and vaccine-injured Americans who complied but now refuse to toe the official line. The hallmark of Lockdown Liberalism is a bullying form of censorship called "cancel culture," which disappears not just the heretical language, but also the heretic who uttered it.

With this letter, I challenge my fellow liberals to look through Galileo's telescope, as it were.

Below, I deconstruct—with scientific citation—the key canons of the reigning liberal mythology and throw down this gauntlet to the liberal intelligentsia to defend their assumptions on the battlefield of scientific debate.

1

Did COVID Vaccines Really Save Millions and End the Pandemic?

With the rising unpopularity of mandates, governments are rushing to declare the pandemic ended, often assigning credit to mass vaccination.[3] However, there is meager scientific evidence that vaccines reduced COVID infections or deaths. To the contrary, there is abundant evidence that mass vaccination had only very brief efficacy against COVID, including the now-undeniable fact, summarized in the February issue of the *European Journal of Epidemiology*, that "Countries with a higher percentage of population fully vaccinated have higher COVID-19 cases per 1 million people."[4]

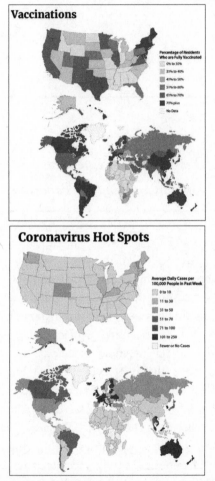

Data based on CDC COVID Tracker and the *New York Times* Interactive Tracking the Coronavirus.[5,6]

Consistent with this global pattern, US deaths attributed to COVID in 2022 were—after mass vaccination—higher than they were in 2020, before vaccination.[7] Aegon Insurance reported a 2021 third-quarter rise of 40 percent in US COVID-19 deaths among people under 65 years old, "the highest percentage in any quarter since the pandemic began."[8] In March 2022, South Korea, one of the most vaccinated nations on Earth, reported record-high COVID infections and mortalities following its aggressive national booster program.[9] COVID deaths in March in Korea exceeded all prior fatalities combined. Likewise, Australia, another mass vaccination leader, saw record-breaking COVID-19 outbreaks in 2022 with deaths 1,700 percent higher than at the start of the pandemic.[10] The tendency of COVID vaccinations to *increase* COVID illness and mortality is a predictable outcome of the well-documented phenomenon of vaccine-induced "pathogenic priming," which I describe in Section 4 below.

Despite the global propaganda effort to persuade us otherwise, the experience of Korea and Australia is the norm. The two-minute video of Johns Hopkins data graphically shows that mass vaccination demonstrated "negative efficacy" against infection (in other words, cases or deaths were higher in the vaccinated

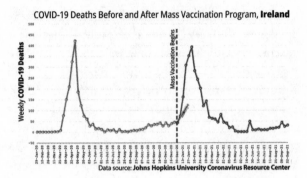

COVID-19 Deaths Before and After Mass Vaccination Program, **Ireland**

Data source: **Johns Hopkins University Coronavirus Resource Center**

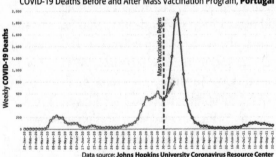

COVID-19 Deaths Before and After Mass Vaccination Program, **Portugal**

Data source: **Johns Hopkins University Coronavirus Resource Center**

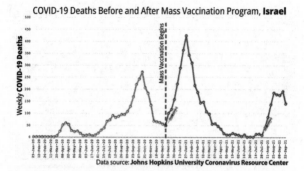

COVID-19 Deaths Before and After Mass Vaccination Program, **Israel**

Data source: **Johns Hopkins University Coronavirus Resource Center**

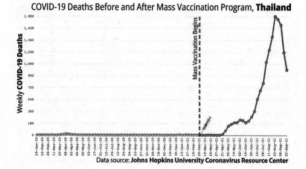

COVID-19 Deaths Before and After Mass Vaccination Program, **Thailand**

Data source: **Johns Hopkins University Coronavirus Resource Center**

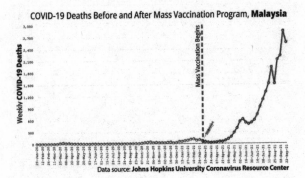

COVID-19 Deaths Before and After Mass Vaccination Program, **Malaysia**

Data source: **Johns Hopkins University Coronavirus Resource Center**

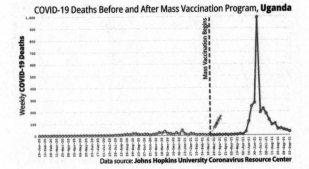

COVID-19 Deaths Before and After Mass Vaccination Program, **Uganda**

Data source: **Johns Hopkins University Coronavirus Resource Center**

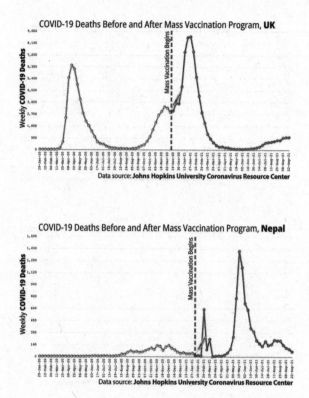

COVID-19 Deaths Before and After Mass Vaccination Program, **UK**

Data source: **Johns Hopkins University Coronavirus Resource Center**

COVID-19 Deaths Before and After Mass Vaccination Program, **Nepal**

Data source: **Johns Hopkins University Coronavirus Resource Center**

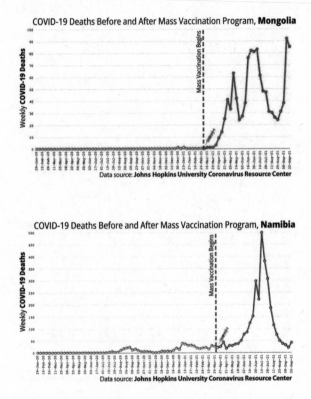

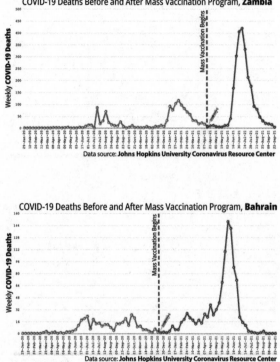

COVID-19 Deaths Before and After Mass Vaccination Program, **Zambia**

Data source: **Johns Hopkins University Coronavirus Resource Center**

COVID-19 Deaths Before and After Mass Vaccination Program, **Bahrain**

Data source: **Johns Hopkins University Coronavirus Resource Center**

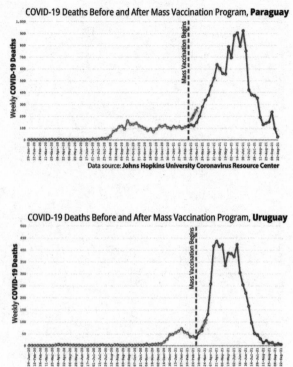

COVID-19 Deaths Before and After Mass Vaccination Program, **Paraguay**

Weekly COVID-19 Deaths

Mass Vaccination Begins

Data source: **Johns Hopkins University Coronavirus Resource Center**

COVID-19 Deaths Before and After Mass Vaccination Program, **Uruguay**

Weekly COVID-19 Deaths

Mass Vaccination Begins

Data source: **Johns Hopkins University Coronavirus Resource Center**

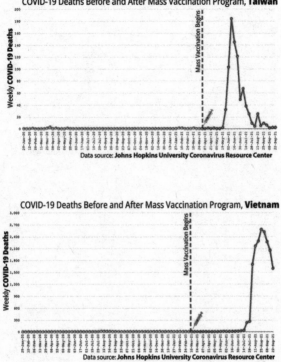

COVID-19 Deaths Before and After Mass Vaccination Program, **Taiwan**

Data source: **Johns Hopkins University Coronavirus Resource Center**

COVID-19 Deaths Before and After Mass Vaccination Program, **Vietnam**

Data source: **Johns Hopkins University Coronavirus Resource Center**

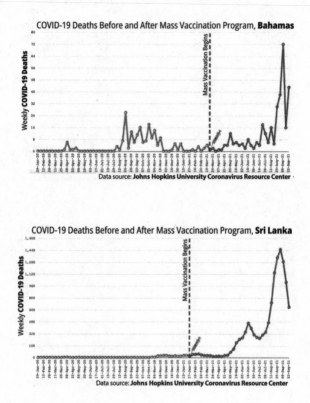

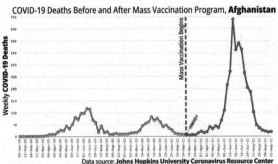

COVID-19 Deaths Before and After Mass Vaccination Program, **Afghanistan**

Data source: **Johns Hopkins University Coronavirus Resource Center**

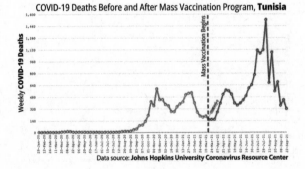

COVID-19 Deaths Before and After Mass Vaccination Program, **Tunisia**

Data source: **Johns Hopkins University Coronavirus Resource Center**

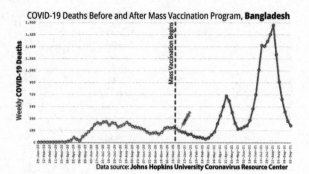

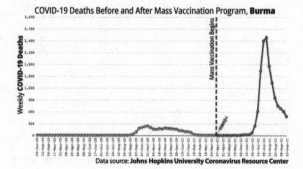

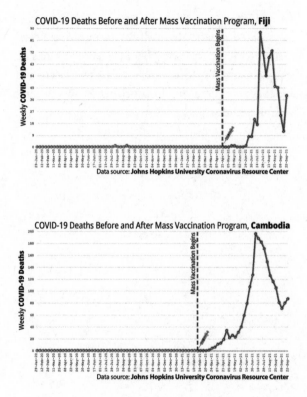

than in the unvaccinated), with COVID-19 cases increasing dramatically in all 145 nations that experimented with the strategy.[11] Because this truth has not been reported by corporate media, it's understandable that you might find it surprising or unbelievable. And, nonetheless, it's true.

2

*Why Were the Lowest COVID Death
Rates in Countries and States That Relied
on Therapeutic Drugs like Ivermectin and
Hydroxychloroquine, and in Countries with
the LOWEST Vaccination Rates?*

Many of these countries with the lowest COVID
death rates had minuscule vaccine coverage. Haiti,
for example, had one of the world's lowest vaccination rates—only 1.4 percent of Haitians got the
jab—and one of the world's lowest death rates from
COVID. According to World Health Organization
data, Haiti suffered only 837 deaths from a population of 11,681,526.[12] Likewise, Nigeria, with a

1.5-percent vaccination coverage[13] for a single jab, experienced a rate of 15.25 COVID deaths per million population compared to the US death rates nearly 200x higher—2,995 deaths per million.[14] Nigeria provided ivermectin and hydroxychloroquine to the vast majority of its people, while US government officials crusaded to block access to these proven prophylactics. By following Dr. Fauci's protocols, America achieved the world's 16th-worst record in deaths per million population. The US with its single-minded vaccination strategy also racked up the highest overall COVID body count; with only 4.25 percent of world population,[15] the United States endured 16 percent of global COVID deaths.[16] Dr. Fauci's policies yielded fatality rates 63 percent above the average among all industrialized nations.[17,18]

For how much longer can liberal Democrats continue to present this cataclysm as a success story and Dr. Fauci as their medical hero? In contrast, the Indian state of Uttar Pradesh (estimated population 235 million) effectively abolished the pandemic overnight by scuttling Dr. Fauci's protocols and distributing ivermectin and other treatment to its citizens.[19,20,21,22] With only 20 percent of adults fully vaccinated, Uttar Pradesh, which is near the bottom of global COVID immunization rankings, had a COVID rate of 100 deaths per million population.[23,24] Other nations like Japan (233.19 deaths per million) and Singapore (234.09 deaths per million)[25] all ended their pandemics after providing their citizens with ivermectin and/or hydroxychloroquine (or chloroquine).[26,27] Leading front-line physicians like cardiologist Dr. Peter McCullough—the most published physician in the history of his subspecialty—and Dr. Robert Malone,[28] a Pentagon advisor and one of the key developers of the mRNA vaccine technology, and Yale statistician Dr. Harvey Risch, MD, PhD, all say that Dr. Fauci's protocols unnecessarily killed 500,000 to 800,000 Americans.[29,30] Over 100 peer-reviewed studies of ivermectin and hydroxychloroquine support this claim.[31,32] Instead of engaging with the mountainous archives of peer-reviewed science supporting the efficacy of hydroxychloroquine and ivermectin, Pharma-funded mainstream media outlets focused on a single study completed by an investigator with strong financial ties to Bill Gates and his foundation.[33] That study did not find the astonishing benefits from ivermectin against infections, hospitalizations, and death that are otherwise practically unanimous throughout the rich scientific literature. Among the many fatal flaws of this paper, researchers administered IVM for only three days when the minimum effective dose is five.[34]

As I document in my book, *The Real Anthony Fauci*, a little-known federal law makes it illegal for the government to grant Emergency Use Authorization (EUA) to new vaccines when any existing drugs—approved for any purpose—are shown effective against the target disease.[35] My book chronicles how this law forced Dr. Fauci, Bill Gates, and their Pharma associates into a reckless crusade to sabotage ivermectin, hydroxychloroquine, and many other scientifically proven, effective early treatments against COVID-19 in order to clear the path for Big Pharma's lucrative "vaccines only" strategy.

3

*Contrary to Official Promises, the Vaccines
Did Not Prevent Infection or Transmission*

You might recall that government officials like Dr.
Anthony Fauci and influential medical "experts" like
Bill Gates initially sold us the "vaccine-only" solu-
tion to COVID by claiming that the vaccine would
immunize against infection and prevent transmission, thereby ending the pandemic. By June 2022
when Anthony Fauci caught COVID-19—following his fourth vaccination—officials had long since
dropped their once implacable claim that COVID-19
vaccines would prevent COVID-19.

However, some authorities continued to suggest
that the jabs could reduce COVID-19's spread. They

should have known better. Instagram deplatformed
me for pointing out that the vaccine industry's mon-
key studies—in May of 2020—made these claims
doubtful; vaccinated monkeys both caught and trans-
mitted COVID with the same frequency as unvaccinated primates.[36,37] The real-world human data have
since forced even the vaccines' most avid promoters
to admit that their initial claims were false. This short
video shows Dr. Fauci, Dr. Rochelle Walensky, Bill
Gates, and other leading promoters adamantly insist-
ing that the vaccines will prevent infection, transmission, and death and end the pandemic. You will then
hear these same trusted authorities gradually shift the
goalposts, finally admitting that vaccines can prevent
neither COVID nor its spread.[38] We liberals need to ask
ourselves this question: If the vaccinated are equally
likely to spread COVID as the unvaccinated—as Dr.
Fauci now acknowledges—then on what basis do we
justify the draconian mandates that denied unvaccinated workers their jobs, children their education, and
encouraged the bullying and bigotry that made the
unvaccinated reviled second-class citizens?[39,40]

  The irrational stigmatizing and outright bigotry
is real and global.[41] Today, American hospitals rou-
tinely deny lifesaving care to Americans based upon
their vaccination status.[42,43] In June 2022, for exam-

ple, a Vanderbilt University Hospital heart transplant surgeon refused to add a mortally ill six-month-old baby boy named August to the transplant recipient list because the parents declined to fully vaccinate him, including the experimental COVID-19 shot, based on medical and religious grounds. An agreement was eventually reached after much public outcry, and now baby August is currently on the transplant list to receive a heart.[44] Should not such a baseless act of brutality against infants offend every liberal conscience?

Leading liberals endorse the exclusion of unvaccinated people from civil rights, including jobs, education, and transportation. The ACLU has called for expanded censorship to silence government critics (see David Cole in the *New York Times*).[45] Even the left's iconic guru Noam Chomsky has recommended the exclusion of the unvaccinated from society. As to food, he says, "Well, that's actually their problem."[46]

Australia is only one of multiple countries that confine unvaccinated, infected, and exposed citizens to internment camps.[47,48] Meanwhile, Austria adopted a slightly more humane option of confining unvaccinated citizens to house arrest.[49,50] Governments across Asia, Africa, and the world banned protests and dissent, authorized extreme forms of oppression, and jailed thousands.[51,52] Rasmussen poll data show widespread liberal support for confining the unvaccinated and taking away their children.[53]

Because of the officially enforced bigotry against unvaccinated people—and considering the long history of invoking pretense of contagious disease infectivity[54,55] to justify segregation laws against despised minorities—don't we as liberals have a special duty to independently investigate government claims that unvaccinated people are more likely to spread contagions? Consider the April 19, 2022, Gridiron Dinner in Washington, D.C. Seventy-two prominent partygoers, who all had to show proof of vaccination, tested positive for COVID after the event.[56] Both Dr. Fauci and Dr. Walensky attended the party.[57] That same week, COVID-19 infected some 200 passengers during a Carnival Cruise Line voyage on which all passengers and crew members were fully vaccinated.[58] Shouldn't the Gridiron and Carnival Cruise superspreader events have made clear to everyone that the vaccines prevent neither cases nor spread, as acknowledged by these public health leaders in August 2021?[59]

The directors of both the CDC and the NIAID acknowledge that the dubious supposition that the unvaccinated are more likely to be silent spreaders is unsupported by science.[60]

4

In Contrast to Official Claims That They Are "Safe and Effective," COVID Vaccines Appear to Show "Negative Efficacy"—Making the Vaccinated More Susceptible to COVID

Most alarming are recent data from Israel,[61] Sweden,[62] the UK's National Health Service database,[63] the New Zealand Ministry of Health database,[64] and Qatar.[65] Among these is a large, peer-reviewed Swedish study published June 2022 in *The Lancet*, which confirmed that vaccine efficacy drops into negative territory 7 to 9 months after vaccination. This means the vaccine is actively contributing to more infections.

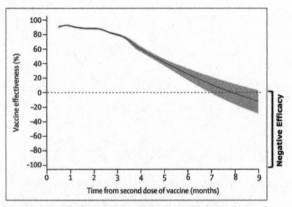

Figure 2: Vaccine effectiveness (any vaccine) against SARS-CoV-2 infection of any severity in 842 974 vaccinated individuals matched to an equal number of unvaccinated individuals for up to 9 months of follow-up
The association is shown using proportional hazards models with 95% CIs (shaded areas) and restricted cubic splines. The model was adjusted for age, baseline date, sex, homemaker service, place of birth, education, and comorbidities at baseline.

These studies and the New York State Health Department's database[66] corroborate the early fears—initially voiced by Dr. Fauci[67] and by leading vaccine developers and promoters including Dr. Peter Hotez[68] and Dr. Paul Offit[69]—that improperly tested COVID vaccines could do permanent damage to the human immune system,[70] paradoxically raising the risks of infection and death from COVID.

In March 2020, Dr. Fauci[71] (along with Drs. Hotez and Offit) warned—based on extensive historical experience with experimental coronavirus vaccines—that COVID jabs, through the mechanism of "pathogenic priming" (also known as Antibody-Dependent Enhancement [ADE]) might make vaccinated individuals *more* susceptible to COVID rather than less. True to these predictions, the COVID vaccine benefits appear to wane rapidly, drifting across the threshold into negative efficacy after about 6 months.[72] This means that vaccinated individuals then become more likely to suffer from COVID infections, hospitalizations, and deaths than unvaccinated individuals.

An analysis of Moderna's randomized controlled trial published in May 2022 based on joint research from scientists at a number of prestigious health institutions shows that mRNA vaccines may actually impair the immune system's ability to fight COVID-19 long term.[73,74] Post-mass vaccination data from government databases around the world support this troubling finding that vaccinated individuals are more likely to become COVID infected.

New York State's vast vaccine database (365,502 children) shows that among children 5 to 11, Pfizer's mRNA vaccine has a mere 12-percent efficacy for one month after kids were "fully vaccinated."[75,76] Then, five weeks after becoming "fully vaccinated," this age group is 40 percent *more* likely to be COVID infected than those children who never received mRNA shots.[77] To confirm this astonishing finding, see Figure 2 below.

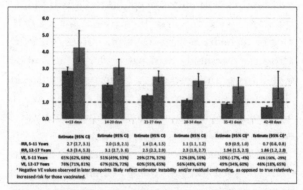

	Estimate (95% CI)	Estimate (95% CI)	Estimate (95% CI)	Estimate (95% CI)	Estimate (95% CI)*	Estimate (95% CI)*
IRR, 5–11 Years	2.7 (2.7, 3.1)	2.0 (1.9, 2.1)	1.4 (1.4, 1.5)	1.1 (1.1, 1.2)	0.9 (0.9, 1.0)	0.7 (0.6, 0.8)
IRR, 12-17 Years	4.3 (3.4, 5.3)	3.1 (2.7, 3.6)	2.5 (2.2, 2.9)	2.3 (1.9, 2.7)	1.94 (1.5, 2.5)	1.86 (1.2, 2.8)
VE, 5–11 Years	65% (62%, 68%)	51% (49%, 53%)	29% (27%, 32%)	12% (8%, 16%)	-10% (-17%, -4%)	-41% (-66%, -29%)
VE, 12-17 Years	76% (71%, 81%)	67% (62%, 72%)	60% (55%, 65%)	56% (48%, 63%)	49% (34%, 60%)	46% (18%, 65%)

* Negative VE values observed in later timepoints likely reflect estimator instability and/or residual confounding, as opposed to true relatively-increased risk for those vaccinated.

Figure 2: Incidence rate ratios, comparing cases during January 3–January 30, 2022, for unvaccinated versus children newly fully-vaccinated December 13, 2021–January 2, 2022, by time since full vaccination.

Within this figure, the dark gray represent 5–11-year-olds, whereas the light gray bars represent 12–17-year-olds. The y-axis is the relative incidence of COVID-19 in the unvaccinated compared to the vaccinated. The x-axis shows bars by increasing time intervals from the date of vaccination. Just after vaccination (i.e., less than 13 days), vaccine immunity peaks, and, for a brief period, unvaccinated 5–11-year-olds are almost three times more likely to get COVID-19 compared to vaccinated, and unvaccinated 12–17-year-olds are over four times more likely to get COVID-19 compared to vaccinated. However, immunity then wanes quickly and precipitously (shown by the consistent decrease in dark and light gray bars). By 42 to 48 days, vaccinated 5–11-year-olds are actually *more likely to contract COVID-19* compared to the unvaccinated. Of course, you have not read anything about this in corporate media reports, which is precisely why your first reaction might be to disbelieve the facts.

UK data similarly show a dramatic rise in COVID susceptibility in vaccinated versus unvaccinated cohorts. After a six-month honeymoon, vaccinated individuals suffered increased COVID risk compared to unvaccinated in every age group.[78,79,80] On March 20, 2022, the UK National Health Service stopped publishing these data after risks of COVID infections among vaccinated individuals in some age groups climbed past 300 percent over unvaccinated.[81] Similarly, both CDC and the *New York Times* suspiciously stopped publishing daily postings of the vaxxed vs. unvaxxed case and death comparisons in March and April 2022, when the graphs began to show no benefit from vaccination.[82,83,84] The CDC has since released its unsubstantiated claim that there were twice as many new cases in the unvaccinated group in May 2022. CDC simultaneously promised to supply data to support this assertion, and has thus far failed to publish those data.

New Zealand Health data show the triple vaccinated are more vulnerable to COVID infection and hospitalization than the unvaccinated.[85] Israeli data also show disproportionate COVID infections among vaccinated.[86] A study of 100,000 Qataris published in the New England Journal of Medicine on June 15, 2022, found that individuals vaccinated with two doses of either the Pfizer or Moderna vaccines were more likely to contract Omicron than unvaccinated. Both vaccines dropped to negative efficacy six months after the second injection— Pfizer dropped to –3.4 percent, and Moderna dropped to –10.3 percent.[87, 88]

A British study published in Science in June 2022 sought to explain why vaccinated individuals are so much more susceptible to infection than the unvaccinated.[89] The paper concludes that the vaccines alter the body's all-important T-cell immunity by making T-cells hyper-vigilant toward the original—now extinct—Wuhan version of COVID-19 while diminishing their capacity to combat new variants like Omicron. Former New York Times reporter Alex

Berenson summarizes the findings in the Science paper thus:

> In other words, the mRNA shots appear to permanently wrongfoot the immune systems of people who receive them and bias them toward producing T-cells to attack variants that no longer exist, even though they never were infected with those variants at all.[90]

Vaccinologists call this phenomenon "original antigenic sin (OAS)." Multiple sources have now confirmed OAS to be a serious problem for people who have taken COVID-19 vaccines.[91]

This is what you see when you look in the telescope and follow the science: government mandates of shoddily tested, heavily subsidized, rushed, zero- liability vaccines that are causing more harm than good.

* * *

In June 2022, I laid out all of the available science and data that proves vaccinating children for COVID is not only unnecessary but will recklessly endanger their lives. You can read that detailed description here: "RFK, Jr's letter to FDA VRBPAC members," ChildrensHD.org/Letter-FDA.

5

The Most Reliable Data Suggest That COVID Vaccines Do Not Lower Risk of Death and Hospitalization

After having to relinquish all their earlier claims to efficacy against both infection and transmission, COVID vaccine promoters have more recently swiveled to advance the dubious premise that COVID vaccines at least reduce the risk of death and hospitalization. For example, this article and this billboard repeat the common propaganda trope that unvaccinated are 16x more likely to die.[92] And here is Sanjay Gupta parroting the official—and scientifically baseless—government mantra that COVID vaccines are "very close to 100 percent in terms of preventing deaths."[93] This claim—that vaccines are effective against serious illness and death—appears to be a final, anemic redoubt for defending the global mass-vaccination enterprise. However, the peer-reviewed published literature from countries with the best data systems and the most widespread vaccine coverage explodes this assertion as yet another bait and switch. This study from Israel—the earliest and among the most vaccinated nations—shows 90.48 percent of hospitalized COVID patients fully vaccinated in July.[94] Some 78 percent of Israelis were then fully vaccinated.[95] This means the vaccinated were actually more likely to end up hospitalized than the unvaccinated.

Similarly, this February 24, 2022, UK Government (NHS) study, summarized in this article, shows 90 percent of hospitalized British COVID patients are fully vaccinated, with four out of five of them triple vaccinated.[96,97] At that time, 70.8 percent of Brits aged 18 and over had received at least three doses.[98] Furthermore, since the start of the Omicron wave in December 2021, overall death rates in England were higher in those doubly vaccinated six months prior than the unvaccinated.[99]

Likewise, government data from February 2022 from New South Wales, Australia, and summarized in this article show vaccinated make up 87 percent of cases, 87 percent of hospitalizations, and 77 percent of deaths among Australians.[100,101] By February 21, 2022, some 80.17 percent of Australians were vaccinated.[102]

Similarly, Ontario COVID death data show that Canadians who received a booster shot have a 50-percent greater death rate than those who received only two vaccines.[103]

Many people will be understandably surprised or even resistant to acknowledging these realities. They are, nonetheless, realities. The failure of the mainstream media to report the truth does not change the truth.

Comparisons of COVID death rates among more or less vaccinated nations also suggest that the vaccines do not prevent COVID deaths. For example, comparing two neighboring Caribbean nations, Haiti with 1-percent vaccination coverage suffered only 835 COVID deaths,[104,105] while Cuba with its slightly smaller population and 88 percent fully vaccinated suffered a COVID death rate 10 times greater—8,529.[106,107]

US media recently published a spate of articles reporting accounts from doctors that dying patients in American hospital COVID wards are disproportionately unvaccinated.[108] However, few, if any, of these reports are either peer-reviewed or published in scientific journals. They are, instead, anecdotal observations by COVID ward doctors. Their conclusions are likely artifacts of CDC's deceptive scheme to classify all patients as "unvaccinated" if two weeks have not elapsed since their second vaccine injection. Furthermore, following CDC recommendations, hospitals classify all patients as "unvaccinated" who fail to report vaccine status upon hospital admission (if, for example, they are distressed, confused, or unconscious). Nurses and doctors on closed COVID wards seeing "unvaccinated" designations on the charts of intubated patients may naturally assume that these are accurate descriptions. In any case, it is difficult to reconcile any anecdotal claim against hard data from other countries showing the opposite results for the exact same vaccines.

6

High Injury and Death Rates from COVID Vaccines May Cancel Out Even the Most Exaggerated Claims of Vaccine Efficacy

A May 2022 preprint in the scientific journal *The Lancet* made the sobering finding that while vaccines may slightly reduce COVID deaths—at least in the short term—they still produce no mortality advantage.[109] FDA and CDC's voluntary injury surveillance system lends credence to this finding.

This revelation should have been fatal to the mass vaccination roll-out and should have immediately ended all government and private efforts to compel vaccination. But then, a month later, in June 2022, another devastating Lancet preprint concluded that mRNA vaccines are causing more serious injuries and hospitalizations than they are averting: "The excess risk of serious adverse events of special interest [from mRNA vaccines] surpassed the risk reduction for COVID-19 hospitalization relative to the placebo group for both Pfizer and Moderna vaccines." The authors of this study include some of the most esteemed eminences in the scientific pantheon including UCLA's iconic statistician and epidemiologist Dr. Sander Greenland, Dr. Patrick Whelan from UCLA, and The BMJ senior editor Dr. Peter Doshi, and others. These scientists based their devastating conclusions on a meticulous reanalysis of the clinical trial data for those two vaccines. In any rational universe, this paper would ring the death knell for mRNA vaccines and trigger an immediate moratorium on public health recommendations for these products.[110]

The wave of vaccine carnage chronicled by the government's own database should have alerted public health regulators to these dangers long before the publication of these troubling articles. Over the 17 months following the rollout launch, CDC's Vaccine Adverse Event Reporting System (VAERS), a voluntary reporting system used by doctors, nurses, hospitals, and individuals, has recorded an astonishing 1,329,135 adverse events,[111] of which 241,910 have been designated by the agencies as serious injuries[112] (including a long list of cardiac, neurological, autoimmune, and reproductive system diseases, blood clots, strokes, myocarditis, seizures, paralysis, hepatitis,[113,114] demyelinating diseases, Guillain-Barré syndrome, Bell's palsy, herpes, vaginal lesions, diabetes, spontaneous abortions, heart attacks—including among six-year-olds, genital ulcers—even in girls as young as five,[115] and other devastating harms).

VAERS Update (through 7/1/2022) of COVID-19 Vaccine Reported Adverse Events:

Condition Reported	Number of Reported Cases	*Estimated Total Caused by COVID Vaccines
Ischaemic stroke[116]	2,084	64,604
Haemorrhagic stroke[117]	320	9,920
Myocarditis[118]	15,120	468,720
Seizures[119]	9,936	308,016
Paralysis[120]	2,154	66,774
Demyelination[121]	282	8,742
Guillain-Barré Syndrome[122]	2,770	85,870

Condition Reported	Number of Reported Cases	*Estimated Total Caused by COVID Vaccines
Bell's Palsy[123]	6,533	202,523
Herpes simplex[124]	529	16,399
Vaginal lesions[125]	20	620
Diabetes[126]	817	25,327
Myocardial infarction[127]	4,521	140,151
Hepatitis[128]	416	12,896

*Underreporting factor of 31 based on Rose 2021. *Science, Public Health Policy, and the Law* 3:100-129[129] by Brian S. Hooker, PhD, Children's Health Defense.

Recent studies from Israel[130] and Taiwan[131] link COVID vaccination with outbreaks of shingles in young people (36–61 age range). Researchers posit that the mRNA vaccine triggers the reactivation of the nascent herpes zoster virus. Pfizer lists shingles as one of the 1,291 adverse events of special interest in its safety document released March 2022 by the FDA under a FOIA court order.[132]

Looking at these alarming data sets, it's not surprising that UK government health data for May 2022 show that five times as many vaccinated individuals are hospitalized compared to unvaccinated individuals for non-COVID injuries and illnesses.[133,134] A May

2022 study by one of Europe's largest university hospitals suggests that serious injuries occur in 8 of every 1,000 vaccines—a rate 40x higher than the government claimed.[135]

VAERS data suggest that the vaccinated are dying at unprecedented rates. VAERS has recorded 29,273 reported deaths since the vaccine rollout began.[136] That is more fatalities over eighteen months than have been reported for all the billions of vaccines put together during the 32 years since CDC and FDA created their reporting system.

These horrifying numbers undoubtedly underestimate the actual casualties; HHS's own study indicates that VAERS captures "fewer than 1 percent of

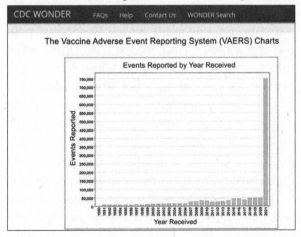

actual injuries."[137] Unfortunately, CDC has resisted 20 years of warnings by public health officials and advocates that it needs to implement a more accurate injury surveillance system, as the law requires.[138]

A March 13 report of national data analyzed by a German health insurer estimates that 2.5 to 3 million Germans have required medical treatment for COVID vaccine injuries.[139] Credible estimates from numerous databases suggest between 150,000[140] and 388,000[141] US deaths since January 2021 are linked to vaccination.[142,143] You have likely been told nothing about any of these alarming findings by major media outlets.

The vaccine's meager efficacy and the avalanche of vaccine-induced injuries and deaths were predictable—indeed, predicted: Pfizer's damning summaries of its own six-month clinical trial data—which the company submitted to FDA to win licensure—showed that 22,000 vaccines must be administered to prevent a single death from COVID and revealed that vaccinated individuals in Pfizer's trial had a 23-percent increased death risk from all causes over six months and a 400-percent elevated risk from fatal cardiac arrest over the unvaccinated cohort.[144] *Pfizer's own data, therefore, suggests that for every COVID death that the vaccine averts, it will, over time, kill four additional people from cardiac arrest.*

As senior editor of *The BMJ*, Peter Doshi, has pointed out, the pitifully small clinical trial population is too tiny to make any reliable predictions about safety and efficacy.[145] But it was Pfizer's choice to limit the study population, and Pfizer is therefore stuck with these disastrous implications.

Even worse, captive FDA regulators allowed Pfizer to unblind (and therefore end) its projected four-year safety study after FDA had received only a median two months of data on trial subjects.[146] This appalling act of pharmaceutical fraud and public health malpractice should outrage even the most entrenched Lockdown Liberals. More alarming still is the current public health objective of collectively inoculating the last remaining Americans, including children and toddlers, with the defective, dangerous, and ineffective jab, a strategy that appears intended to eliminate the global control group.

7

Mass Vaccination Has Preceded Global Rises in Excess Death

In 1976, CDC pulled the swine flu vaccine after 25 reported deaths.[147] In contrast, CDC and its slavish and scientifically illiterate media allies have responded to the tsunami of fatalities and injuries linked to COVID jabs by hiding harms from the public. Nevertheless, insurance companies are reporting massive waves of unexplained excess deaths (increases in all-cause mortality) in 2021—among previously healthy vaccinated Americans. Practically all the increases occur in younger ages. According to J. Scott Davison, the CEO of OneAmerica, a national life insurance corporation headquartered in Indiana, excess deaths are up 40 percent in the third quarter of 2021.[148] These deaths are primarily non-COVID deaths among workers aged 18 through 64. "We are seeing right now the highest death rates we have ever seen in the history of this business. Previous crises pale in comparison to the pandemic," Davison told journalists in December. "A one-in-200-year catastrophe would be a 10-percent increase over prepandemic [levels]. So 40 percent is just unheard of."[149]

Even more alarming, Lincoln National, the nation's fifth largest insurance company, reported a 163 percent increase in death benefit payouts from group life insurance policies during 2022 from "non-pandemic-related mortality." The company paid out $500 million in 2019 prior to the COVID pandemic, $548 million in 2020 during the height of the COVID pandemic, and a stunning $1.4 billion in 2021 during the national mass vaccination crusade. The company also reported an astonishing rise in total claims, including deaths and injuries of $6 billion for the year. The company paid out $23 billion in 2019, the baseline year, $22 billion in 2020, the pandemic year, and $28 billion in 2022, the year of mass vaccination.[150] The latest data from the Insurance Regulatory and Development Authority of India have reported similar numbers.[151]

While COVID-19 killed the infirm and elderly, many at the end of life, COVID vaccines are associated with deaths among the young and fit. Former Blackrock Portfolio Manager Ed Dowd reports an analysis of CDC's mortality and morbidity data by insurance industry actuaries showing a shocking 85-percent rise in excess mortality in the millennial age group—ages 25 to 44—the worst-ever increase in history. Says Dowd, "The millennial age group saw 61,000 deaths, excess deaths, in a one-year time frame with an acceleration into the mandates and the boosters in the fall. I mean, they just experienced a Vietnam War in a year."[152] Death-certificate data from the CDC similarly show a 40-percent increase in excess mortality of 18-to-49-year-olds during a 12-month period ending in October of 2021, with only 42 percent of those deaths attributed to COVID.[153] Most alarmingly, the military's health database—famous for its precision and rigor—has reported a 1,100 percent increase in mortality and morbidity among the 18–49 age group of vaccinated military personnel in 2021 over previous years.[154,155] Shouldn't liberals be alarmed that service members reported more deaths to VAERS (as suspected vaccine casualties) than COVID deaths?[156] UK National Health Service data show the COVID-19 death rate for children up to 19 years old tripled after vaccination. Prior to mass vaccination, the death rate was 2.6 children per month. Post mass vaccination, deaths have increased to 7.8 children per month.[157]

Public-health officials and Pharma's media allies have reacted by ignoring the waves of sudden unexplained deaths among those of working age[158] and disregarding the astonishing frequency with which highly conditioned athletes are collapsing right on the fields of play. As of April 26, 2022, the website GoodScience has chronicled 992 athlete cardiac arrests with 644 dead following COVID shots.[159] Schools are normalizing the idea of heart attacks[160] in children. Israel has begun equipping public schools with defibrillators,[161] and some US schools are now commonly requiring children to submit to cardiac testing in order to qualify for school sports.[162]

As part of its campaign to normalize this epidemic of sudden unexplained deaths in healthy young people, CDC has created a new cause of death, "Sudden Adult Death Syndrome" (SADS). Liberals should be alarmed at the ease with which CDC persuaded our bought and brain-dead media that "SADS is all a big mystery."[163, 164]

A May 5, 2022, Israeli article published in Nature documented a 25 percent increase over 2019–2020 numbers in ambulance calls responding to cardiac arrests and acute coronary syndrome in the 16 to 39-year-old population. Associated with rates of the first and second vaccine doses in this age group. The study found that the cardiac and coronary emergencies were not associated with COVID-19 infections.[165, 166]

An April 2022 study of 23 million people in *JAMA Cardiology* found an elevated myocarditis risk of almost 700 percent following the Moderna vaccine.[167] And sadly, myocarditis cases suffer 50-percent mortality within five years, so we have not seen the end of these harms.[168] Finally, a November 2021 German study found that the German states with the highest vaccination rates are the ones suffering the highest excess mortalities.[169] A July 2022 study by New Zealand's University of Waikato links COVID booster shots to sixteen excess deaths for every 100,000 doses.[170]

I realize that for many people, the first instinct is to resist taking these realities onboard, to assume that everything I have shared here (with citations) must be untrue. Resistance to information is often the signal that reading it again, with the willingness to actually check the original sources, is called for. If even a fraction of what I have shared here is true, and you acknowledge that you haven't been told this truth till now, that itself should be alarming.

8

Pharma and CDC Have Hidden the Damning Injury and Death Data Reports with Cooperation from Leading Media Outlets Including the New York Times

Government transparency and pugnacious press skepticism are hallmark liberal values, and yet, the *New York Times*, in February, made the belated—and apparently, untroubled—admission that the CDC has been systematically withholding data that challenge its official narratives and cherry-picking data to promote the government/Pharma "safe and effective" orthodoxy.[171] Furthermore, both Pfizer and FDA petitioned a federal court to deny the public access to Pfizer's clinical trial data for the next 75 years.[172] So much for the promised transparency! These are the same data that FDA reviewed for only 108 days prior to giving Pfizer its license. Immediately afterward, two top vaccine regulators at FDA resigned.[173] Is there any good reason that these systematic concealments—strikingly reminiscent of the industry/FDA collusion and deception that precipitated the opioid crisis—should not trigger liberal indignation?

9

Should Government Technocrats Be Partnering with Media and Social Media Titans to Censor and Suppress the Questioning of Government Policies?

Most of my fellow liberals are unaware of all these alarming facts due to a highly orchestrated global pandemic of journalistic malpractice. At the outset of the pandemic, most of the world's leading news organizations—BBC, Reuters, AP, AFD, CBC, CNN, CBS, ABC, *Washington Post*, *Financial Times*, Facebook, Google/YouTube, Microsoft, Twitter, and others—organized themselves into a collusive anti-democratic and anticompetitive cartel known as the

Trusted News Initiative (TNI)—pledged to squelch and censor all reports about government COVID countermeasures that challenged official proclamations.[174] The link you just passed goes to the official BBC website, where you can read, in the words of those conspirators, about how a group of companies that have historically competed against one another to reveal government untruths have now partnered in lockstep to promote the Government line and connived to work in concert to attack reporting that runs counter to officially proclaimed orthodoxies. These organizations have successfully prevented virtually all honest journalism about vaccine injuries and vaccine failure from reaching the general public.

This rigid compliance with Big Pharma's propaganda agenda was, unfortunately, no great leap for mainstream media. In recent years, Pharma and its allies have made enormous investments to control American newsrooms and transform mainstream and social media, TV networks, and scientific journals into vessels for mercantile propaganda. The pharmaceutical industry is now the dominant advertiser on television, the funding source for over 75 percent of total advertising and an even greater percentage during news shows.[175] Furthermore, Pharma investor Bill Gates has

distributed $319 million in recent years to news orga-
nizations specifically targeting "independent" plat-
forms like NPR, Public Television, *The Independent*,
The Guardian, etc., that were historically less susceptible to pressure from commercial advertisers.[176] Most
alarming, since the pandemic's outset, HHS has quietly paid out over a billion dollars (you read that right)
to news outlets like CNN, the *Washington Post*, and
the *New York Times* to promote COVID vaccines.[177]
Those companies have obligingly published thousands
of pieces extolling vaccination while actively censor-
ing criticism of vaccines—or content that challenges
Pharma profit taking—all without disclosing those
compromising payoffs to their audiences.

With this cash in hand, the US media have
abandoned their traditional skepticism toward government edicts and abetted the censorship of non-
conforming views. While the media abolish contrary
opinion, they marginalize, vilify, and bully dissent-
ers. In a recent editorial, Johns Hopkins professor Dr.
Marty Makary, MD, MPH, author of *The Price We
Pay: What Broke American Health Care—And How
to Fix It*, writes: "Throughout the pandemic, *The
New York Times* and other outlets have only sourced
doctors on the establishment groupthink bandwagon,
dangled fear to young people and blindly amplified

every edict government doctors fed without asking questions, just as the press did with weapons of mass destruction in Iraq."[178,179]

When will liberals face the fact that scientific fact is not always the self-interested pronouncements of Sanjay Gupta or Anthony Fauci? The current head of the CDC should not represent the beginning and end of scientific argument. As Galileo understood, monolithic orthodoxies are the enemies of science. Science is a dynamic and continuous search for empirical truth, and the term "scientific consensus" is therefore an oxymoron. Second opinions are critical safeguards of both patient care and public health. I challenge liberals to, at long last, engage with the many esteemed and principled scientists and physicians and other experts who have presented well-evidenced and principled opposition to the official dogma promoted by Big Pharma and its government allies. Liberals must once again practice the sifting and winnowing of truth from falsehood that is the hallmark of the liberal tradition—and to stop outsourcing this crucial task to health-care bureaucrats, scientifically illiterate, deadline-harried journalists, celebrity doctors, and mercenary fact-checking organizations financed by Pfizer, Johnson & Johnson, and vaccine investors like Bill Gates and Mark Zuckerberg.

10

Virtually All My Early Predictions Have Matured from "Conspiracy Theories" to Proven Facts

Anthony Fauci is not the only celebrity to continue to tout COVID vaccines from a sickbed. It's common to hear double or triple-vaccinated—or even quadruple-vaccinated (like Dr. Fauci)—liberal icons like Nancy Pelosi, Jimmy Kimmel, Bill Gates, Peter Hotez, Jimmy Fallon, Whoopi Goldberg, Joy Behar, Kelly Ripa, Amy Schumer, Stephen Colbert, Miley Cyrus, Hillary Clinton, Sean Penn, Barack Obama, Piers Morgan, Elton John, Tori Spelling, James Corden, Hoda Kotb, Seth Meyers, Jen Psaki, Elizabeth Warren, Cory Booker, and Kamala Harris continue to enthusiastically defend COVID vaccines even after their second and even third bout of post-vaccine COVID. Would these individuals remain so blithely satisfied if three or four polio vaccines failed to protect them against polio? These true believers seem immune to logic, reason, or evidence. Shouldn't the massive and undeniable vaccine failures—in the wake of all those extravagant promises about their efficacy—and the constantly shifting narratives raise liberal ire and skepticism instead of blind allegiance to a failed product? Two years ago, the mainstream media were disparaging as "dangerous crackpots" or "conspiracy theorists" anyone who questioned mandates for masks, social distancing, lockdowns, or the accuracy of PCR tests, and anyone who suggested that COVID-19 might have originated in the Wuhan lab or who pointed out that natural immunity[180] was superior to vaccine-induced immunity, or that children had low risk for COVID and no reason to risk vaccination, or that mRNA vaccines might alter human DNA. But, in the course of time, all these "conspiracies" have proven true.

10.1) Masks Are Ineffective and Dangerous

In recent weeks, with over 100 studies now attesting that masks do not stop viral spread and over

60 studies showing they can cause physical and mental health injuries and developmental delays, the CDC has finally and quietly acknowledged that mandates for cloth masks make little scientific sense.[181,182] A comprehensive April 2022 study of masking practices in 35 European countries found no benefits from masking in preventing COVID-19 disease or death. Instead, masking showed a highly significant correlation with increased risk of COVID death.[183,184] In the same month, another study found that lung tissue of British citizens was saturated with microplastics, likely an artifact of two years of mask compliance.[185]

A preprint study[186] by Chandra and Høeg replicated the methodology used earlier by CDC[187] but extended CDC's analysis to a much broader, nationally diverse population over a longer interval. It found no difference in pediatric COVID case rates between school districts that mandated masks versus those that did not.

Even the *New York Times* now acknowledges that "Covid has spread at a similar rate as in mask-resistant cities. Mask mandates in schools also seem to have done little to reduce the spread. Hong Kong, despite almost universal mask-wearing, recently endured one of the world's worst Covid outbreaks."[188]

10.2) Social Distancing Was Not Science-Based

CDC also dropped its social-distancing guidelines as former FDA Commissioner Dr. Scott Gottlieb confessed publicly that the six-foot rule his agency implemented was "arbitrary and not science-based."[189,190]

10.3) School Closures Were Not Science-Based

Former CDC Director Robert Redfield has, likewise, confirmed that the school closures the CDC recommended had no scientific basis.[191] Furthermore, there was no effort by government authorities to understand the devastating collateral damage of school lockdowns to children. The predictable damage to a generation of children from the closures has proven cataclysmic.

A 2020 report by the esteemed education research consortium NWEA predicted that even the 2020 school year's relatively short two-month lockdown would irretrievably damage US students.

> Preliminary COVID slide estimates suggest students will return in fall 2020 with roughly 70% of the learning gains in reading relative to a typical school year. However, in mathematics, students are likely to show much smaller learning gains, returning with less than 50% of the

learning gains and in some grades, nearly a full year behind what we would observe in normal conditions.[192]

The European Commission report entitled "The likely impact of COVID-19 on education" using information from international datasets found that poor children would suffer the most grievous deficits:

> . . . a reduction in scores of between 6.5 and 14 points. The switch from offline to online learning caused by COVID-19 is expected [. . .] to exacerbate existing educational inequalities. More vulnerable students, such as for instance those from less advantaged backgrounds, are especially likely to fall behind during this emergency period. These students are less likely to have access to relevant learning digital resources (e.g. laptop/computer, broadband internet connection) and less likely to have a suitable home learning environment (e.g. a quiet place to study or their own desk). Additionally, they may not receive as much (direct or indirect) support from their parents as their more advantaged counterparts do.[193]

Dr. Fauci did not seem to comprehend that education deficits pose a devastating public health risk. Lost education, for example, affects life expectancy. A *JAMA* article predicts that the school closures for 24.2 million US schoolchildren will result in the loss of 13.8 million years of life.[194] No one has yet calculated the global loss of life from school closures that occurred when most other nations followed the US protocols in lockstep. We do know that the lockdown response interrupted over a billion children's schooling, leaving millions never to return.[195,196]

Numerous studies indicate that the academic declines from school closures are accompanied by alarming mental health injuries.

A report by Collateral Global[197] found that eight out of ten UK children and adolescents suffered an increase in anxiety, loneliness, and stress, with one in six children reporting "significant mental health problems,"[198,199] and one in four feeling "unable to cope" during the lockdowns.[200] Eighty percent of young people reported a "deterioration in their emotional well-being,"[201] and local public health institutions saw dramatic increases in self-harm and eating disorders, along with an "explosion" of children with disabling tic disorders,[202] and "record numbers of children being prescribed antidepressants,"[203] with the intensity of these negative feelings correlating to the duration of school closures.

A damning report by UNESCO and then jointly reissued with UNICEF and the World Bank reported that school closings disproportionately affected the world's poor.[204] "Classroom closures continue to affect more than 635 million children globally, with younger and more marginalized children facing the greatest loss in learning after almost two years of Covid."[205]

The UNESCO report predicted that the percentage of ten-year-olds in low- and middle-income countries who cannot read or understand a simple text will rise to 70 percent.

The *Guardian* reports that a quarter of the world's school systems are on the verge of collapse:

> As much of the developing world faces a combination of interrelated crises including extreme poverty . . . there are growing fears for a 'lost generation of learners.'[206]

The losses in academic competence will translate into future lower earnings for the student cohorts directly affected by the lockdown—primarily in poorer countries with less resilience.

10.4) Lockdowns Were Counterproductive

Mass lockdowns of the healthy contradicted a century of public health practice including the WHO's official guidelines[207] and contradicted the frantic protests of over 600 scientists who addressed a letter to President Trump on May 19, 2020, warning that lockdowns would create an economic and public health catastrophe far more damaging than COVID.[208] To date, some 76,000 PhD scientists, physicians, and others have signed the Great Barrington Declaration, a statement originally drafted by three of the world's top statisticians from Harvard, Stanford, and Oxford Universities disputing the public health efficacy of lockdowns, and recommending, instead, the long-accepted pandemic protocol: quarantining the sick, and targeted protection of the vulnerable.[209]

As predicted, global lockdowns have cost a cataclysmic $16 trillion, according to the International Monetary Fund.[210] An extensive meta-analysis of dozens of peer-reviewed studies by Johns Hopkins researchers has confirmed that while they devastated global economies, the Trump/Biden lockdowns did practically nothing to reduce the spread of COVID.[211] All they did was prolong the pandemic and amplify its pain. In March 2022, the most wide-ranging study on COVID restrictions to date found that the states with the most stringent lockdowns fared far worse (New York, California, New Jersey, and Illinois got F grades). In stark contrast, states that allowed their citizens more freedoms (Florida, Utah, Nebraska, Vermont, Montana) fared far better.[212] Even Dr. Fauci has recently confessed that scientific evidence does not support the supposition that lockdowns were effective.[213] In this April 2022 TV interview, Dr. Fauci finally acknowledged his true strategy behind lockdown mandates—a psychological warfare technique to coerce vaccine compliance: "You use lockdowns to get people vaccinated."[214]

10.5) Vaccinating Children Causes More Harm and Death Than It Averts

On March 10, 2022, CDC admitted, in response to a Freedom of Information request, that it has not a single record of a healthy child under age 15 dying from COVID.[215] Comprehensive research from Germany[216] and the UK,[217] and separate studies of US children by Johns Hopkins,[218] *Nature*,[219] and *The Lancet*,[220] had all previously reaffirmed that healthy children have statistically zero risk of dying from COVID. Meanwhile, the vaccines impose a high risk (1/2,700) of causing myocarditis in 12–17-year-old boys.[221,222] A grim analysis of recent United

Kingdom Office for National Statistics (ONS) data from January 1, 2021, through January 31, 2022, analyzed by Dr. Wayne Winston, PhD, professor emeritus of Decision Sciences at the University of Indiana's Kelley School of Business, suggests that vaccinated children are more likely to die from any cause than unvaccinated children. The data show that vaccinated children ages 10 to 14 are 28 times more likely to die than unvaccinated, and the vaccinated 15-to-18-year-olds are 1.82 times more likely to die than unvaccinated teens of the same age.[223]

Though you have not read about it, four Scandinavian countries have banned the Moderna vaccine in people below age 30 because myocarditis is killing more of them than COVID.[224] By lying about the seizures that confined a 13-year-old volunteer, Maddie de Garay, to a wheelchair and feeding tube in its *New England Journal of Medicine* report on the adolescent trial, Pfizer also deceived regulators and the public about the 1/1,300 occurrences of devastating neurological injuries in 12-to-15-year-old girls during Pfizer's clinical trial. Pfizer reported Maddie's injury as "stomachache."[225] This was only one of many glaring irregularities and outright frauds that Pfizer perpetrated to win fast-track licensure.[226]

10.6) Officials Wrongly Used PCR Tests to Justify the Countermeasures

After imposing masks, social distancing, and lockdowns on the basis of PCR tests, Dr. Fauci finally admitted on December 29, 2021, that the PCR was inadequate for detecting actual COVID infections: "PCR doesn't measure replication competent viruses . . . it doesn't give you any indication of whether or not you're transmissible."[227,228,229] Imagine that: The method that millions of people have been using to determine if they have a transmittable infection "doesn't give you any indication of whether or not you're transmissible."

As government officials now admit, the PCR tests were identifying many false positives from previous infections (and from other sources), thus creating a kind of pandemic echo effect where one case multiplied itself through time. Government officials mandated cataclysmic policies based on false perceptions created by ghost combatants.

The CDC similarly used faulty PCRs to inflate death rates from COVID. The CDC recommended that hospitals classify every death as a COVID casualty so long as the patient, at some point, produced a positive PCR. HHS incentivized[230] this recommendation with unspeakably generous financial bonuses to the hospitals.[231, 232] The CDC now acknowledges that only 6 percent of COVID casualties are certain to have died from COVID. The remaining 94 percent suffered from an average of 3.8 potentially lethal comorbidities, any of which could have been the true cause of death.[233]

And you likely did not know that.

10.7) COVID-19 May Have Come from Wuhan Lab

US intelligence analysts now suggest that a lab leak from the Wuhan Institute of Virology is a plausible origin for the virus.[234] A March *Vanity Fair* investigation suggests Dr. Fauci misled two presidents and orchestrated a global cover-up to deceive the world about COVID-19's origins.[235]

10.8) Natural Immunity is Superior to Vaccine Immunity

The CDC now acknowledges that natural immunity is superior to vaccine-induced immunity.[236] An Israeli study shows that natural immunity is 27 times more durable and provides a broader spectrum of protection against a wide range of variants than inoculations![237] Anthony Fauci nevertheless continues to promote vaccinating previously infected children, a practice he has for 40 years emphatically condemned.[238]

10.9) The Weight of the Science Suggests That COVID mRNA Vaccines Can Indeed Alter Human DNA

While the CDC continues to reassure us on its web page that "The genetic material delivered by mRNA vaccines never enters the nucleus of your cells," Swedish researchers at Lund University on February 25, 2022, published a study showing that messenger RNA (mRNA) from Pfizer's COVID-19 vaccines rapidly enters human liver cells and reverse-transcribes into DNA.[239,240] The terrifying ramifications of these findings are difficult to overstate. They raise the real possibility that the mRNA vaccine may permanently alter the human genome, potentially priming—from mothers to children—a DNA code that might continually produce spike proteins known to damage nervous systems, brain, bone marrow, and immune systems and to produce blood clots. In April 2022, CDC finally admitted that it had no data to support its claim that mRNA vaccines don't alter DNA.[241]

All this suggests that, during the first two years of the COVID crisis, I and Children's Health Defense and other government policy critics like Brownstone Institute, The Frontline Critical Care Doctors, Dr. Mercola, Green Med Info, and numerous other critics of the reigning orthodoxy have been far more reliable sources of accurate vaccine information than the public health authorities and the media. The CDC, FDA, and NIH have been the leading promoters of a tsunami of "vaccine misinformation."

11

Instead of a Public Health Response, Dr. Fauci's Militarized and Monetized COVID Policies Proved to Be a Devastating War on the Poor, Children, and the Working Classes

Orchestrated fear and blind trust in Dr. Fauci and entrenched opposition to Donald Trump caused many liberals to abandon not only their antipathy for censorship, but the other core concepts of liberal ideology that had been our proud legacy since 1932: a solicitude for the poor, workers, minorities, and vulnerable children.

Anthony Fauci's quarantine was a prolonged pajama party for upper-crust Americans who could afford DoorDash food deliveries and Amazon shopping. Lockdowns provided a novel adventure in telecommuting for the laptop elites, and a cushy year of remote education for their children. But even mainstream critics are increasingly recognizing that COVID policies have devastated workers, the poor, minorities, and children. The lockdowns created nearly five hundred new billionaires—who are now gorging on the bleached cadavers of America's massacred middle-class—and engineered a $3.98 trillion shift in wealth from the poor and working Americans to a new oligarchy of Pharma billionaires, social and mainstream media titans, surveillance state robber barons, and military contractors.[242,243] Shouldn't liberals be particularly skeptical that media and social media billionaires raked in fortunes from the lockdowns while actively collaborating with government officials to censor and deplatform critics of these controversial and failed policies?

Government countermeasures were, in contrast, a nightmare for the middle class and the poor. Shouldn't liberals worry that the lockdowns greatly increased inequality between a rich controlling few and a rapidly expanding disempowered poor, reversing years of poverty reduction that liberal administrations have fought for since FDR's New Deal?[244, 245, 246]

The number of active business owners in the United States plummeted by 3.3 million from February to April 2020, crushing the backbone of democratic capitalism.[247] The lockdown's leading corporate promoters, Amazon, Facebook, Microsoft, Google, and Walmart, flourished while their Main Street competitors withered and died.

As I touched on above, multiple studies show that lockdowns, school closures, masking,[248] and childhood COVID vaccines[249] had catastrophic impacts on children and the poor with casualties and costs far exceeding the deaths and injuries from COVID-19.[250] Globally, lockdowns pushed over 130 million people into food insecurity and caused millions of deaths from starvation.[251,252] A UNICEF report estimates that 60 million ADDITIONAL children will grow up in poverty and malnutrition.[253] Millions of girls have been forced into child marriage.[254] Reduced case-finding and treatment access for tuberculosis and HIV/AIDS has left more infected people untreated, to transmit to others and die.[255,256] The World Health Organization reports that over 52,000 additional children under five died from malaria in the WHO African Region in 2020 alone.[257] Another UNICEF report estimates that lockdowns are responsible for the deaths of hundreds of thousands of children—228,000 in South Asia alone.[258,259] In July 2020, the Associated Press reported that US and European lockdowns were starving to death 10,000 African kids each month.[260]

The International Finance Facility (IFF) considers that twice as many children died from lockdowns as died from COVID-19.[261] Lockdown-related mortalities and morbidities are likely to outlast COVID countermeasures. The Bank of International Settlements, key to international finance, recognizes that gross domestic product is a major determinant of long-term health, and the economic carnage—including ruinous national debt, mass bankruptcies, and inflation—from global lockdowns are likely to hobble the worldwide economy—and therefore, public health—for decades.[262]

COVID countermeasures seemed to disproportionately injure minorities. As discussed in Section 2 above, the death rate from COVID-19 in Nigeria—which had a vaccination rate of 1.3 percent, was 15 deaths per million population—about 1/200 the death rate among Americans of 3800 per million population. Haiti, with a vaccination rate of 1.4 percent, had a COVID death rate of less than 14 per million population.[263] Why is it that COVID-19 barely touched scantily vaccinated African and Haitian populations while American Blacks died at 3.6x and Latinos died at 2.5x the rates of Whites in all age categories?[264] Life expectancy among American Blacks dropped by 3.25 years.[265] Shouldn't this be a subject of intense curiosity and energetic inquiry? Yet, Dr. Fauci has never acknowledged, much less attempted to explain these troubling discrepancies.

Lockdowns also disproportionately harmed children, increasing child labor, teenage pregnancy rates, and child marriages.[266] US maternal mortality reached record highs during the pandemic, with the highest deaths in Black women.[267,268] While health authorities closed public schools, police padlocked playgrounds and basketball courts in minority neighborhoods. Children skipped school lunches that, for many, were their sole nutritious meal. A heartbreaking 24 percent of American teens report experiencing hunger.[269] The lockdowns diminished eye health due to increased screen time and aggravated the obesity epidemic. Even children became more obese during the pandemic.[270] Worsening an existing obesity crisis—Americans gained 29 pounds on average in 2020 to 2021—may have contributed to high death counts since obesity was and is a leading COVID comorbidity.[271,272] The lockdowns, themselves, were counterproductive since regulators knew early in the pandemic that COVID spread indoors and not outdoors.[273] Lockdowns aggravated not just obesity, but stress and vitamin D deficiencies, all of which proved deadly COVID comorbidities.

The only indicator of poverty and social deterioration that seemingly improved during the quarantine was child abuse. Sadly, this was an artifact of less reporting. Most child abuse reports emanate from schools, and the reported incidents naturally diminished when schools closed. While schools officially stopped reporting suspected child abuse, our government health panjandrums locked abused children indoors with their abusers. A recent CDC summary disclosed a shocking 55.1 percent of teens regularly experienced emotional abuse during lockdowns, and 11.3 percent, physical abuse—up from 13.9 percent and 5.5 percent, respectively, in 2013.[274]

Dr. Fauci acknowledges that he never considered the amplifying impacts of lockdowns on collateral damage including aggravating existing epidemics of isolation, mental health, and obesity.[275,276,277] UNESCO reports that child and youth mental health has become a crisis within a crisis. Children globally experienced social isolation, disruption to daily routines, stress associated with parental unemployment, and feelings of uncertainty about their future.[278] The CDC reports that between

January and June of 2021, almost 20 percent of US teen-agers contemplated suicide, 9 percent attempted to kill themselves, and 44.2 percent reported feeling persistently sad and hopeless.[279,280] Suicide is now the second-leading cause of death for Black children, and it's rising still.[281] The poor disproportionately shouldered terrible increases in alcoholism, drug addiction, overdoses, retarded development, and mental illness. Gun sales and carjacking have hit record highs. CDC reports dramatic upticks in shootings and other violent crime including a 30 percent rise in homicides, pedestrian deaths, reckless driving (despite dramatic reductions in miles driven), disorderly and disruptive unhinged behavior, anger, and fits of anger in schools and society at large.[282,283]

The lockdowns didn't just worsen mental health and obesity. They lowered IQ and impacted early childhood developmental milestones in infants and toddlers. Masks, for example, impaired emerging speech and language skills in children during critical developmental stages, while lockdowns deprived children of important growth stimuli.[284,285] Infants born during the pandemic are missing speech development milestones normal for babies their age. Researchers warn that due to lockdowns and other disruptions, nearly one-third of elementary students will need "intensive support to become proficient readers."[286,287]

Children born during the pandemic are at a greater risk for academic failure because parents haven't been able to engage their babies and toddlers in the types of conversations that are "crucial for language development."[288] On average, American toddlers lost an astonishing 22 IQ points during the lockdowns according to a Brown University study.[289]

That longitudinal observational study also found that the poor and minority children shouldered the heaviest burden of lost functionality. The evidence from the US is particularly clear:

> We find that children born during the pandemic have significantly reduced verbal, motor, and overall cognitive performance compared to children born pre-pandemic. Moreover, we find that males and children in lower socioeconomic families have been most affected.[290]

A study in *JAMA* recorded the dramatic losses in developmental skills:

> Compared with the historical cohort, infants born during the pandemic had significantly lower scores on gross motor, fine motor, and personal-social skills.[291]

In France, a study entitled "Adverse Collateral Effects of COVID-19 Public Health Restrictions on Physical Fitness and Cognitive Ability" found:

> An alarming decline in both overall physical fitness and cognitive performance among primary school French children due to the public health restrictions imposed in order to slow down the spread of the COVID-19 virus.[292]

An intractably corrupt CDC's response has been to normalize the injuries by readjusting childhood milestones. Under CDC's revised milestones, adopted last month, a "normal" child will now be expected to walk at 18 months rather than 12 and is expected to have learned 50 words at 30 months rather than 24 months.[293,294] I know that reading all this is a massive dose of new truth and reality for most people. Rather than resisting the information, readers might ask why they haven't been told all this by their trusted news sources.

Today, the last people in America still required to cover their faces are children, minimum wage earners, waiters, waitresses, servers, staff, and front-line "essential workers," who risk their lives to deliver meals and Amazon packages to the more privileged classes. For many of them, the masks have become potent symbols of orchestrated fear, obedience, subservience, and the dehumanizing anonymity of being poor and powerless.

Contemporary liberalism has endeavored to veil its abdication of its traditional role as a defender of minorities and the poor with a symbolic embrace of BLM iconography, and by deploying the new liberal weapon of censorship to cancel racist speech. Such symbolic gestures—such as posting a black square—are anemic substitutes for genuine empowerment for minorities and the downtrodden. Lip service has replaced action, and the strategies liberals have embraced badly undermine their expressed intentions.

Finally, a Harvard study by David Cutler and Lawrence Summers estimates the cost of the pandemic to the US government at $16 trillion.[295] This ruinous debt is yet another devastating attack on our children. That cataclysmic expenditure—intended to benefit fragile elderly in the last years of their lives (and there is little evidence that it did)—did so by beggaring the young from the coming generations. American children of the future will have less access to health care, food, education, security, and opportunity for home ownership. They will suffer diminished lives, in a weakened, impoverished country. They will also suffer the loss of democracy, civil rights, and power over their lives.

12

America Has Endured an Unprecedented
Attack on Our Bill of Rights

In 2001, liberal Democrats, led by my uncle, Senator Edward Kennedy, mobilized to block the Patriot Act, a 3,000-page assault on the US Constitution drafted by Neocons and stampeded through Congress in the frenzy of post-9/11 hysteria and propaganda. That bill laid the groundwork for the emerging Security State. In contrast, during the COVID crisis, liberal leaders colluded with disgraced and discredited Neocons and public-health technocrats to impose unprecedented infringements on personal liberties and human rights that have dramatically accelerated the rise of the security state, now biosecurity state, while expanding autocratic rule and elevating a new plutocracy.

The recent constitutional infringements included the normalization of censorship, the forced shuttering of churches nationwide, and the curtailments to our rights to gather, protest, and petition—via social distancing, lockdowns, vaccine passports, and mandates—all of which are in violation of the FIRST AMENDMENT's protections of *speech, religious worship,* and *assembly*. Bureaucratic diktats trampled the FIFTH AMENDMENT by shuttering millions of businesses with *neither due process nor just compensation*; abolished the SEVENTH AMENDMENT right to *jury trials* for injuries caused by pharmaceutical companies, doctors, and hospitals; and obliterated the FOURTH AMENDMENT *prohibitions against warrantless searches*. Pandemic countermeasures promoted track-and-trace surveillance and the systematic incursions upon traditional privacy rights including unprecedented collection of private data. Finally, the cascade of technocratic edicts violated the FIFTH and FOURTEENTH AMENDMENT *guarantees of due process by dispensing with notice and comment rulemaking, public hearings, and environmental impact statements* prior to imposition of broad and intrusive public-health fiats by unelected bureaucrats.

The Canadian government pioneered another bold shortcut to our new corporatist dystopia: silencing dissent and obliterating the *right to assemble and petition* by shuttering the bank accounts of peaceful protestors without trial and even without bothering to charge these citizens with any crime.[296] The global juggernaut to replace hard coin and cash with digitalized and programmable currencies will likely make this Canadian innovation the norm in the USA and across the former liberal democracies worldwide. Today, nearly 100 nations are now planning transitions to digital currency.[297]

13

*The Deprivation of Rights and Economic
Demotion of the Middle Class Have Spawned
a Rebellion. Its Ideology Is Fluid and
up for Grabs.*

All the infringements and the cataclysmic economic
demotions of middle-class and poor families have,
predictably, prompted millions of disaffected and
alienated Americans to coalesce into an emerging
rebellion. Working and poor Americans—once the
Democratic Party's core constituencies—are rising to reclaim their rights and livelihoods in a class
war against the globalist elites and corporate titans
who collaborated with the regulatory technocracy to

steal middle-class wealth, property, and jobs and to impose the oppressive mandates without either public hearings or scientific citation. In 1966, my father, Senator Robert Kennedy, predicted that vast discrepancies in wealth and the routine abuse of power by Latin American oligarchs would precipitate revolutions that would be hijacked by communists if the American government continued to ally itself with the military and the oligarchs, instead of the disenfranchised poor and workers:

> A revolution is coming—a revolution which will be peaceful if we are wise enough; compassionate if we care enough; successful if we are fortunate enough—But a revolution which is coming whether we will it or not. We can affect its character; we cannot alter its inevitability.[298]

Liberal media and social media outlets mischaracterize all dissent from official orthodoxies as right-wing Trumpism. The current liberal cosmology incorrectly casts working-class populists—including protesting truckers—as right-wing racist Trump fanatics or as "deplorables." But the populist movement that has coalesced to oppose the mandates is racially and religiously diverse, ideologically incoherent, and, increasingly, sees itself as engaged in a class war against Big Tech, Big Data, Big Pharma, Big Banking, Big Media, and Wall Street titans, who themselves are aligned with police and intelligence, military, and security state forces. Moreover, the liberal response of allying the Democratic Party with Pharma and its captive technocrats and global elites of the "Davos Billionaire's Club" in blind support for mandates plays directly into the hand of right-wing demagogues like Donald Trump.

Liberals are missing a great opportunity here to reenfranchise the working class by acknowledging the inequitable impact of so many of the pandemic policies that Donald Trump and leading Democrats fast-tracked and green-lit with patriotic fervor. The pandemic is an equity issue that goes way beyond the condescension of liberals toward people of color who must, the patronizing liberal mythology holds, be gently disabused of their current vaccine skepticism based on true, but now supposedly irrelevant, horrors of medical experimentation on their "ancestors." The "essential workers" were disproportionately people of color; the people who lost their jobs to the lockdowns were also disproportionately people of color; the people who lost their jobs due to noncompliance with mandates were mostly from the working classes;

and the people who died of COVID, vaxxed or not, were disproportionately people of color. The party is abandoning much of its base for a cabal of elites, represented almost exclusively by journalism, academia, and the Beltway. For its own survival, the left must learn to critically think again.

Does Hatred of Donald Trump and Support for Pfizer's Vaccine Justify Savaging the Constitution?

Today the unifying passions of the Democratic Party are a ferocious hatred of Donald Trump, an orchestrated and often irrational fear of COVID, and blind support for all public-health mandates. Like every hatred, Trumpaphobia gives power to the object of its abomination. Ironically, Trump is now dictating policy choices for Democrats. Liberals turned against the heroic whistleblower Julian Assange when Trump expressed his sympathy for the jailed journalist and free speech and civil rights advocate. Democrats abandoned their sturdy opposition to the Trans-Pacific Partnership when Trump criticized the TPP. Similarly, Trump's occasional and minor disdain for

Dr. Anthony Fauci blinded Democrats to Fauci's five-decade role as the architect of industry capture of the public-health agencies. Trump's endorsement of hydroxychloroquine prompted liberals to toss the remedy into the same dumpster with Trump's global-warming denialism, despite overwhelming scientific support for the remedy.

Trump's cozy relationship with Vladimir Putin is arguably one of the factors that coaxed traditional anti-war liberals into ignoring Ukrainian President Volodymyr Zelensky's anti-democratic track record and his troubling relationship with the openly Nazi and anti-Semitic Azov Battalion and won liberal support for a military intervention that has, predictably, enriched petroleum companies and military contractors, doubled gasoline prices, and ignited the galloping inflation that will further beggar the American middle class.[299] The Ukrainian crisis has pushed liberals—including President Biden—to embrace the monumentally savage and tyrannical Saudi ruler Mohammed bin Salman (MBS)[300] whose naked aggression against Yemen has killed over 377,000 Yemenis—mainly civilians, dwarfing, in its homicidal brutality, the Russian invasion of Ukraine.[301] In October 2018, a year before the COVID pandemic, MBS sent a team of assassins to strangle the Washington Post journalist Jamal Khashoggi and dismember him with a bone saw.[302] Liberals once reviled MBS, and President Biden's recent olive branch to the Crown Prince, has draped liberalism in shame.

In their ardor for demonstrating revulsion of Trump, liberals have walked away from the hallowed core values of liberalism. Traditional FDR/Kennedy liberalism revered civil rights and personal freedoms, including free speech and expression, freedom to assemble and petition, religious freedom, and bodily autonomy. Historically, liberals were champions of labor and the poor, and the enemies of autocrats and bullies. But today . . . Well, not so today.

For decades, liberals once proudly harbored a deep skepticism toward Pharma, and the military-in-dustrial/intelligence complex, and generally nurtured an antipathy for war, a contempt for blind obedience to undeserving authorities, a wariness of officials who used fear as a governing tool (recall FDR's admo-nition that "The only thing we have to fear is fear itself"), and solicitude for bodily autonomy (isn't our mantra "My body, my choice"?).

During the COVID coup d'état, the same liber-als who had fought the Patriot Act and opposed the Iraq War suddenly adopted the Neocons' hostility to the Bill of Rights, their affinity for a national security state, and their embrace of a bellicose, expansionist "regime-change" foreign policy.

In fact, aside from fury toward Trump and a wild love affair with Pharma vaccines, contemporary liberalism's only remaining artifacts of traditional FDR/Kennedy liberalism are its concern for environmental sustainability and its defense of ethnic and LGBTQ minorities from bigotry and official bullying. Liberals seem unaware that their acquiescence to censorship and the erosion of the Bill of Rights to silence opponents of Big Pharma's vaccines will invariably open the door to King Coal, Big Oil, Big Ag, and Big Chemical and all their captive agencies likewise silencing their own inconvenient critics. Military contractors and their government captives are already wielding the new tool of censorship to silence debate on the Ukraine intervention. Both supporters and opponents of that intervention should worry about this development. The weapons that liberals condone to silence Pharma's critics will invariably be deployed against the most vulnerable populations liberals have always championed. That is the inexorable lesson of history.

In the Oscar Award-winning 1966 film, *A Man for All Seasons*, the character of Britain's Chancellor

Sir Thomas More explains why it's unwise to cut down the Constitution even for the well-intentioned objective of destroying the Devil:

> Cut a great road through the law to get after the Devil?... And when the last law [is] down, and the Devil turn[s] 'round on you, where [will] you hide, the laws all being flat? This country is planted thick with laws, from coast to coast, Man's laws, not God's! And if you cut them down..., do you really think you could stand upright in the winds that would blow then? Yes, I'd give the Devil benefit of law, for my own safety's sake!

As they demolish the Constitution to get at Trump and promote Pharma, liberals don't seem to appreciate that the most vulnerable population will soon reap the whirlwind. And where will we all hide when an unleashed Exxon, Smithfield, Peabody, Monsanto, and Koch Energy have license to silence their critics with the help of Google, Facebook, and Twitter? As these soulless multinational behemoths deploy their new powers to silence dissent, to strip-mine our landscapes, pollute our waterways, exterminate our wildlife, and commoditize our children, they are fanning the toxic winds that will soon envelop America in a cytokine storm of environmental desolation and dystopian totalitarianism.

Rather Than Hiding from Debate,
Liberals Should Take Every Opportunity
to Defend Their Strategies before
Bipartisan Audiences

Orchestrated fear, systematic censorship, and fierce tribalism have pushed our country into a political polarization more perilous than at any time since the American Civil War. If we are to find a better future—if we are to imprint this rising dissident movement with liberalism's traditional idealism and avoid its capture by corporate tyrants—liberals need to start talking to people with whom they otherwise disagree.

It is in this spirit that I continue to dialogue with individuals with whom I differ on many other issues. Liberals have criticized and canceled me for speaking with conservative populists like Tucker Carlson and Steve Bannon. If we don't talk to our political opponents, how will we ever find common ground? If we can't explore scientific truths through debate, how will we ever bridge the widening chasm between America's warring tribes?

Those fierce disputes between Republicans and Democrats, Blacks and Whites, vaxxed and unvaxxed serve only the intentions of the global elite who exploit the distraction to systematically rob us of our treasure, our health, and our freedoms.

Coercive policies—especially the suppression of speech and dissent—will only breed more skepticism and mistrust and will never bridge the gap between America's polarized tribes. As Professor Makary observes: "The American people are hungry for honesty. They see the inequity of COVID policies and want the data straight, not politically curated by a small group of like-minded scientists."[303]

Sincerely,

Robert F. Kennedy Jr.

Endnotes

1   Good Jobs First, "Violation Tracker Industry Summary Page," Penalties
    Since 2000, https://violationtracker.goodjobsfirst.org/industry/pharmaceuticals.

2   Amy Mitchell and Mason Walker, "More Americans Now Say Government
    Should Take Steps to Restrict False Information Online than in 2018," Pew
    Research Center, Aug. 18, 2021, https://www.pewresearch.org/fact-tank
    /2021/08/18/more-americans-now-say-government-should-take-steps-to
    -restrict-false-information-online-than-in-2018/?utm_source=Pew
    +Research+Center&utm_campaign=6332d6ef12-EMAIL_CAMPAIGN
    _2021_08_20_03_53&utm_medium=email&utm_term=0_3e953b9b70
    -6332d6ef12-400018517.

3   João Viana et al., "Controlling the Pandemic During the Sars-Cov-2
    Vaccination Rollout," *Nature Communications*, Jun. 16, 2021, https://www
    .nature.com/articles/s41467-021-23938-8.

4   S.V. Subramanian and Akhil Kumar, "Increases in Covid-19 Are Unrelated
    to Levels of Vaccination Across 68 Countries and 2947 Counties in the
    United States," *European Journal of Epidemiology* (2021), https://www.ncbi
    .nlm.nih.gov/labs/pmc/articles/PMC8481107.

5   "Tracking Coronavirus Vaccinations Around the World-'Maps and
    Trackers'," *New York Times*, May 2022, https://www.nytimes.com/interactive
    /2021/world/covid-vaccinations-tracker.html.

6   "CDC COVID Data Tracker," Centers for Disease Control and Prevention,
    May 2022, https://covid.cdc.gov/covid-data-tracker.

7   Centers for Disease Control and Prevention, "Table1. Deaths Involving
    Coronavirus Disease 2019 (Covid-19), Pneumonia, and Influenza Reported
    to NCHS by Time-Period and Jurisdiction of Occurrence," CDC, Data as

of Apr. 26, 2022, https://www.cdc.gov/nchs/nvss/vsrr/covid19/index.htm.
*Search Table 1 at the bottom by Month and Year.

8 Toby Sterling, "Aegon, Other Insurers Hit by U.S. Covid-19 Deaths in Third Quarter," Reuters, Nov. 11, 2021, https://www.reuters.com /business/finance/aegon-q3-operating-result-down-16-us-covid-linked -claims-2021-11-11.

9 TrialSite Staff, "Record Surge of COVID-19 Cases and Deaths in Heavily Vaxxed South Korea—What's Going on?" *TrialSite News*, Mar. 20, 2022, https://trialsitenews.com/record-surge-of-covid-19-cases-and-deaths-in -heavily-vaxxed-south-korea-whats-going-on.

10 "Record Breaking Wave of Covid-19 across Australia Sees Deaths 1700% Higher Than the Start of the Pandemic; & the Fully Vaccinated Account for 4 in Every 5 of Them," *The Exposé*, Apr. 7, 2022, https://dailyexpose.uk /2022/04/07/australia-record-breaking-wave-deaths-4-in-5-vaccinated.

11 Children's Health Defense, "COVID Deaths Before and After Vaccination Programs," *The Defender*, Mar. 3, 2022, https://childrenshealthdefense.org /research-citation/covid-deaths-before-and-after-vaccination-programs.

12 Jim Hoft, "Haiti Did Not Vaccinate Its Citizens, the Current Vax Rate is 1.4%—Yet Country Has One of Lowest COVID Death Rates in the World—Weird, Huh?" *Gateway Pundit*, Jul. 8, 2022, https://www. thegatewaypundit.com/2022/07/haiti-not-vaccinate-citizens-current-vax -rate-1-4-yet-country-one-lowest-covid-death-rates-world-weird-huh/.

13 Olayemi Oladotun, "COVID-19 Vaccination Rates in Nigeria," *Information Nigeria*, Sep. 22, 2021, https://www.informationng.com/2021/09/covid-19 -vaccination-rates-in-nigeria-see-details.html.

14 "Coronavirus (COVID-19) Deaths Worldwide Per One Million Population as of Apr. 26, 2022, by Country," *Statista*, 2022, https://www .statista.com/statistics/1104709/coronavirus-deaths-worldwide-per-million -inhabitants.

15 "United States Population" (Live), *Worldometer* https://www.worldometers. info/world -population/us-population.

16 Worldometer, "Reported Cases and Deaths by Country or Territory," *Worldometer* May 20, 2022, https://www.worldometers.info/coronavirus /#countries.

17 "Why is America's Covid-19 Death Rate So High?" *Advisory Board—Daily Briefing*, Feb. 3, 2022, https://www.advisory.com/daily-briefing/2022/02 /03/us-death-rate.

18 National Center for Health Statistics, "Daily Updates of Totals by Week and State, Provisional Death Counts for Coronavirus Disease 2019

(COVID-19)," CDC, Apr. 15, 2022, https://www.cdc.gov/nchs/nvss/vsrr /covid19/index.htm.

19   "33 Districts in Uttar Pradesh are Now Covid-free: State Govt," *Hindustan Times,* Sep. 10, 2021, https://www.hindustantimes.com/cities/lucknow-news /33-districts-in-uttar-pradesh-are-now-covid-free-state-govt-101631267966925 .html.

20   "Population of Uttar Pradesh," *Census 2021*, https://census-2021.co.in /population-of-uttar-pradesh-2021.

21   Mark Patricks, "Two Heavily Vaccinated Nations, Two Very Different Outcomes," *The League of Power*, Mar. 9, 2022, https://www.lopmatrix. com/two-heavily-vaccinated-nations-two-very-different-outcomes.

22   Maulshree Seth, "Uttar Pradesh Government Says Early Use of Ivermectin Helped to Keep Positivity, Deaths Low," *MSN*, Dec. 5, 2021, https://www .msn.com/en-in/news/other/uttar-pradesh-government-says-early-use-of -ivermectin-helped-to-keep-positivity-deaths-low/ar-BB1gDp5U.

23   Nikhil Rampal, "With Only 20% Adults Fully Vaccinated, Uttar Pradesh at Bottom of Covid Immunisation Rankings," *The Print*, Oct. 26, 2021, https:// theprint.in/health/with-only-20-adults-fully-vaccinated-uttar-pradesh -at-bottom-of-covid-immunisation-rankings/756446.

24   "Mortality Rate Calculator," *Omni Calculator*, https://www.omnicalculator .com/health/mortality-rate?c=USD&v=type:1,deaths:23508,population :235000000.

25   "Coronavirus (COVID-19) Deaths Worldwide Per One Million Population as of Apr. 26, 2022, by Country," *Statista*, 2022, https://www.statista.com /statistics/1104709/coronavirus-deaths-worldwide-per-million-inhabitants.

26   TrialSite News staff, "Chairman of Tokyo Metropolitan Medical Association Declares during Surge, Time for Ivermectin is Now," *TrialSite News*, Aug. 16, 2021, https://trialsitenews.com/chairman-of-tokyo-metropolitan-medical -association-declares-during-surge-time-for-ivermectin-is-now.

27   National Health Commission & State Administration of Traditional Chinese Medicine, "Diagnosis and Treatment Protocol for Novel Coronavirus Pneumonia," Mar. 3, 2020, https://www.chinalawtranslate .com/wp-content/uploads/2020/03/Who-translation.pdf.

28   Harriet Alexander, "The Truth about Joe Rogan's Controversial Guests: 'Father of mRNA' Dr Malone Pointed Out that Hospitals Get Covid Bonuses and Said Biden Government Is 'Out of Control' - While Dr Mccullough Said US Is Hypnotized and 'Pandemic Is Plandemic,'" *Daily Mail*, Feb. 2, 2022, 02:02 EDT, https://www.dailymail.co.uk/news/article -10466285/The-truth-Joe-Rohans-controversial-guests-Dr-Malone-Dr -McCullough.html.

29 "Joe Rogan Interview with Dr. Robert Malone [Transcript and Full Interview]," *Covid Vaccine Side Effects & Information*, Jan. 3, 2022, https://covidvaccinesideeffects.com/joe-rogan-interview-with-dr-robert-malone-transcript.

30 Greg Hunter, "800,000 Lives Could Have Been Saved with Ivermectin and HCQ [Video]" *Shift Frequency*, Mar. 21, 2022, https://www.shiftfrequency.com/800000-lives-could-have-been-saved-with-ivermectin-and-hcq-video.

31 "Ivermectin for COVID-19: Real-Time Meta Analysis of 81 Studies," Version 182, *Ivmeta*, Mar. 21, 2022, https://ivmmeta.com.

32 "HCQ for COVID-19: Real-Time Meta Analysis of 331 Studies," Version 196, *Hcqmeta*, Mar. 23, 2022, https://hcqmeta.com/#fig_fpearly.

33 Gilmar Reis, MD, PhD, Eduardo A.S.M Silva, MD, PhD, Daniela C.M. Silva, et al., "Effect of Early Treatment with Ivermectin among Patients with Covid-19," *NEJM* 386, (2022): 1721-1731, www.doi.org/10.1056/NEJMoa2115869.

34 Pierre Kory, MD, MPA, "Fraudulent Trial on Ivermectin Published by the World's Top Medical Journal. Big Pharma Reigns - Part 2," *Substack*, May 15, 2022, https://pierrekory.substack.com/p/fraudulent-trial-on-ivermectin-published-859.

35 "Emergency Use Authorization of Medical Products and Related Authorities," U.S. Food & Drug Administration, Jan. 2017, https://www.fda.gov/regulatory-information/search-fda-guidance-documents/emergency-use-authorization-medical-products-and-related-authorities#preeua.

36 Neeltje van Doremalen et al., "ChAdOx1 nCoV-19 Vaccination Prevents SARS-CoV-2 Pneumonia in Rhesus Macaques," *bioRxiv*, May 13, 2020, https://www.biorxiv.org/content/10.1101/2020.05.13.093195v1.full.pdf.

37 William A. Haseltine, "Did the Oxford Covid Vaccine Work in Monkeys? Not Really," *Forbes*, May 16, 2020, https://www.forbes.com/sites/williamhaseltine/2020/05/16/did-the-oxford-covid-vaccine-work-in-monkeys-not-really/?sh=3968af163c71.

38 "COVID-19 Vaccine," Vimeo, https://vimeo.com/623540275/a13b682bf.

39 Anika Singanayagam, PhD, Seran Hakki, PhD, Jake Dunning, PhD, et al., "Community Transmission and Viral Load Kinetics of the SARS-CoV-2 Delta (B.1.617.2) Variant in Vaccinated and Unvaccinated Individuals in the UK: A Prospective, Longitudinal, Cohort Study," *The Lancet* 22, no. 2 (2021): 183-195, https://doi.org/10.1016/S1473-3099(21)00648-4.

40 Yahoo Finance, "Dr. Fauci on COVID-19 Spread: Vaccinated People Who Have an... Infection Are Capable of Transmitting," YouTube, Jul. 30, 2021, https://www.youtube.com/watch?v=mP9iHyj1uiU.

41   Divya Bhanot et al., "Stigma and Discrimination During COVID-19 Pandemic," Front Public Health 8, (2020), https://doi.org/10.3389/fpubh.2020.577018.

42   The Associated Press, "Patient Who Refused Covid Vaccine Was Denied a Heart Transplant," *NPR*, Jan. 26, 2022, https://www.npr.org/2022/01/26/1076004339/heart-transplant-patient-unvaccinated.

43   Patty Nieberg, Thomas Peipert, and Colleen Slevin – *The Associated Press*, "Colorado Woman Who Won't Get Vaccinated Denied Transplant," *ABCNews*, Oct. 8, 2021, https://abcnews.go.com/Health/wireStory/colorado-woman-vaccinated-denied-transplant-80467001.

44   Andrew Court, "'Dying' Baby Approved for Heart Transplant after Parents Fight Vaccine 'Mandate'," *New York Post*, Jun. 29, 2020, https://nypost.com/2022/06/29/dying-baby-approved-for-heart-transplant-after-parents-fight-vaccine-mandates.

45   Michael Powell, "Once a Bastion of Free Speech, the A.C.L.U. Faces an Identity Crisis," *New York Times*, Jun. 6, 2021, https://www.nytimes.com/2021/06/06/us/aclu-free-speech.html.

46   *National Post* Staff, "Noam Chomsky Says the Unvaccinated Should Just Remove Themselves from Society," *National Post*, Oct. 27, 2021, https://nationalpost.com/news/world/noam-chomsky-says-the-unvaccinated-should-just-remove-themselves-from-society.

47   Damien Cave, "Australia Is Betting on Remote Quarantine. Here's What I Learned on the Inside," *New York Times*, Aug. 20, 2021, https://www.nytimes.com/2021/08/20/world/australia/howard-springs-quarantine.html.

48   "Inside Australia's COVID Internment Camp," *UnHerd*, Dec. 2. 2021, https://unherd.com/thepost/inside-australias-covid-internment-camp.

49   Kate Connolly, Samantha Lock, and agencies, "Austria to Put Millions of Unvaccinated People in Covid Lockdown," *The Guardian*, Nov. 12, 2021, https://www.theguardian.com/world/2021/nov/12/austria-province-to-place-millions-of-unvaccinated-people-in-covid-lockdown.

50   Chloe Taylor, "Austrian Police Conduct Random Checks to Enforce Covid Lockdowns for the Unvaccinated," *CNBC*, Nov. 17, 2021, https://www.cnbc.com/2021/11/17/austria-covid-lockdown-police-conduct-random-vaccine-status-checks.html.

51   "Covid-19 Triggers Wave of Free Speech Abuse," *Human Right Watch*, Feb. 11, 2021, https://www.hrw.org/news/2021/02/11/covid-19-triggers-wave-free-speech-abuse.

52   "Interactive Map–Covid-19 Triggers Wave of Free Speech Abuse," *Human Right Watch*, https://features.hrw.org/features/features/covid/index.html?#arbitrary.

53 "COVID-19: Democratic Voters Support Harsh Measures against Unvaccinated," Rasmussen Reports, Jan. 13, 2022, https://www .rasmussenreports.com/public_content/politics/partner_surveys/jan_2022 /covid_19_democratic_voters_support_harsh_measures_against _unvaccinated.

54 Edna Bonhomme, "Germany's Anti-vaccination History is Riddled with Anti-Semitism," The Atlantic, May 2, 2021, https://www.theatlantic.com /health/archive/2021/05/anti-vaccination-germany-anti-semitism/618777.

55 Mitchell J. Besser, "Typhus : The Influence of Society and State on a Human Disease Professor," American International Journal of Contemporary Research 5, no. 5 (2016): 107-117, https://www.semanticscholar.org/paper /Typhus-%3A-The-Influence-of-Society-and-State-on-a-Besser/739980d130 9776bbe2e04c44007875d344e7cec9.

56 Christina Zhao and Molly Roecker, "72 People at High-Profile D.C. Dinner Test Positive for Covid," NBC News, Apr. 10, 2022. https://www.nbcnews .com/politics/politics-news/67-attendees-test-positive-covid-high-profile-dc -dinner-rcna23763.

57 Callie Patteson, "COVID Hits More than 70 Attendees of Swanky DC Gridiron Dinner," New York Post, April 11, 2022. https://nypost.com /2022/04/11/covid-hits-more-than-70-attendees-of-dc-gridiron-dinner/.

58 Jack Phillips, "Fully Vaccinated Carnival Cruise Ship Hit with COVID-19 Outbreak," Epoch Times, May 5, 2022. https://www.theepochtimes.com /mkt_app/fully-vaccinated-carnival-cruise-ship-hit-with-covid-19-outbreak _4448297.html.

59 "CDC: COVID Vaccines Won't Stop Transmission; Fully Vaccinated Can Still Get, Spread Delta Strain," The Star Democrat, Aug. 5, 2021, https://www. stardem.com/news/national/cdc-covid-vaccines-won-t-stop-transmission -fully-vaccinated-can-still-get-spread-delta-strain/article_5f83d0cb-8b0a -535d-bbad-3f571754e5ae.html.

60 Heather Chapman, "Fauci: The Vaccinated Can Be as Infectious as the Unvaccinated, so They Need to Wear Masks Indoors; Pandemic Will Get Worse," Kentucky Health News, Aug. 3, 2021, https://ci.uky.edu /kentuckyhealthnews/2021/08/03/fauci-the-vaccinated-can-be-as-infectious -as-the-unvaccinated-so-they-need-to-wear-masks-indoors-pandemic-will -get-worse.

61 Maayan Hoffman, "In Israel, COVID Is No Longer Just a Disease of the Unvaccinated. Why?" Jewish News Syndicate, Feb. 9, 2022, https://www.jns .org/in-israel-covid-is-no-longer-just-a-disease-of-the-unvaccinated-why.

62 Peter Nordström, PhD, Marcel Ballin, MSc, Anna Nordström, PhD, "Risk of Infection, Hospitalisation, and Death Up To 9 Months After a Second Dose of Covid-19 Vaccine: A Retrospective, Total Population Cohort Study

in Sweden," *The Lancet* 399, no. 10327 (2022): 814-823, https://doi.org/10.1016/S0140-6736(22)00089-7.

63 "SARS-CoV-2 Variants of Concern and Variants Under Investigation in England, Technical Briefing 17," *Public Health England*, Jun. 25, 2021, https://assets.publishing.service.gov.uk/government/uploads/system/uploads/attachment_data/file/1001354/Variants_of_Concern_VOC_Technical_Briefing_17.pdf.

64 Guy Hatchard, PhD, "NZ Ministry of Health Data Shows Triple Vaccinated Are Now More Vulnerable to Covid Infection and Hospitalisation than the Unvaccinated," *The Exposé*, Apr. 6, 2022, https://dailyexpose.uk/2022/04/06/nz-moh-data-triple-vaccinated-most-vulnerable-covid.

65 Heba N. Altarawneh, MD, Hiam Chermaitelly, PhD, Hussein H. Ayoub, PhD, et al., "Effects of Previous Infection and Vaccination on Symptomatic Omicron Infections," NEJM, (2022), https://www.doi.org/10.1056/NEJMoa2203965.

66 Vajeera Dorabawila, "Effectiveness of the BNT162b2 Vaccine among Children 5-11 and 12-17 Years in New York after the Emergence of the Omicron Variant," *medRxiv*, Feb. 28, 2022, https://www.medrxiv.org/content/10.1101/2022.02.25.22271454v1.full.pdf+html#page.

67 Chan Zuckerberg Initiative, "Q&A with Mark Zuckerberg and Dr. Anthony S. Fauci, Director of the NIAID," YouTube, Mar. 26, 2020, 00:24:31 - 00:26:25, https://youtu.be/971QcDEha5I.

68 House Committee on Science, Space and Technology, "Coronaviruses: Understanding the Spread of Infectious Diseases and Mobilizing Innovative Solutions," YouTube, Mar. 5, 2020, 00:39:18-00:42:53, https://www.youtube.com/watch?v=8F1L_Mw-DiE.

69 ZDoggMD, "COVID-19: Is Our Cure Worse Than the Disease? With Dr. Paul Offit," YouTube, Mar. 17, 2020, 00:39:02-00:39:28, https://www.youtube.com/watch?v=MAeaFoSuJho&t=1s.

70 Stephanie Seneff, Greg Nigh, Anthony M. Kyriakopoulos, Peter A. McCullough, "Innate Immune Suppression by SARS-CoV-2 mRNA Vaccinations: The role of G-quadruplexes, Exosomes, and MicroRNAs," *Food and Chemical Toxicology* 164, 2022: 113008, https://doi.org/10.1016/j.fct.2022.113008.

71 Chan Zuckerberg Initiative, "Q&A with Mark Zuckerberg and Dr. Anthony S. Fauci, Director of the NIAID," YouTube, Mar. 26, 2020, 00:24:58, https://youtu.be/971QcDEha5I.

72 "Antibody-dependent Enhancement (ADE) and Vaccines," *Children's Hospital of Philadelphia*, Reviewed on Sep. 9, 2021, https://www.chop.edu/centers-programs/vaccine-education-center/vaccine-safety/antibody-dependent-enhancement-and-vaccines.

73   Dean Follmann, Holly E. Janes, Olive D. Buhule, et al., "Anti-nucleocapsid
     Antibodies Following SARS-CoV-2 Infection in the Blinded Phase of the
     mRNA-1273 Covid-19 Vaccine Efficacy Clinical Trial," medRxiv, Apr. 19,
     2022, https://www.medrxiv.org/content/10.1101/2022.04.18.22271936v1.

74   Madhava Setty, MD, "Did Moderna Trial Data Predict 'Pandemic of the
     Vaccinated?'" The Defender, May 4, 2022, https://childrenshealthdefense.org
     /defender/moderna-trial-data-pandemic-vaccinated.

75   Molly Walker, "COVID Vaccine Less Effective in Younger Kids," *MedPage
     Today*, Feb. 28, 2022, https://www.medpagetoday.com
     /infectiousdisease/covid19vaccine/97425?xid=nl_covidupdate_2022-03
     -01&eun=g797189d0r&utm_source=Sailthru&utm_medium=email&utm
     _campaign=DailyUpdate_030122&utm_term=NL_Gen_Int_Daily_News
     _Update_active.

76   Vajeera Dorabawila, Dina Hoefer, Ursula Bauer, et al., "Effectiveness of the
     BNT162b2 Vaccine among Children 5-11 and 12-17 Years in New York
     after the Emergence of the Omicron Variant," *medRxiv*, Feb. 28, 2022,
     https://doi.org/10.1101/2022.02.25.22271454.

77   Alex Berenson, "URGENT: mRNA Shots Raise the Risk of Covid
     Infection in Children Under 12," *Unreported Truths*, Feb. 28, 2022, https://
     alexberenson.substack.com/p/urgent-mrna-shots-raise-the-risk/comments.

78   UK Health Security Agency, "COVID-19 Vaccine Surveillance Report,
     Week 8," Feb. 24, 2022, see Figure 1A, B and C, https://assets.publishing.
     service.gov.uk/government/uploads/system/uploads/attachment_data/file
     /1057599/Vaccine_surveillance_report_-_week-8.pdf.

79   Amanuensis, "Infection Rates Higher in Triple Vaccinated Than in
     Unvaccinated across All Age Groups, UKHSA Data Show," *The Daily
     Sceptic*, Mar. 13, 2022, https://dailysceptic.org/2022/03/13/infection-rates
     -higher-in-triple-vaccinated-than-unvaccinated-across-all-age-groups-ukhsa
     -data-show.

80   "COVID-19 Vaccine Surveillance Report, Week 10," *UK Health Security
     Agency*, Mar. 10, 2022, https://assets.publishing.service.gov.uk/government
     /uploads/system/uploads/attachment_data/file/1060787/Vaccine_surveillance
     report-_week_10.pdf.

81   Amanuensis, "Vaccine Effectiveness Hits as Low as Minus 300% – as
     UKHSA Announces It Will No Longer Publish the Data," *The Daily Sceptic*,
     Mar. 20, 2022, https://dailysceptic.org/2022/03/20/vaccine-effectiveness
     -hits-as-low-as-minus-300-as-ukhsa-announces-it-will-no-longer-publish
     -the-data.

82   Meryl Nass, MD, "CDC and NY Times Stopped Revealing the Vaxxed vs.
     Unvaxxed Case and Death Comparisons 6 and 8 Weeks Ago, When the
     Graphs Began to Show No Benefit from Vaccination/ NY Times," *Anthrax*

Vaccine – Posts by Meryl Nass, MD, Jun. 1, 2022, http://anthraxvaccine
.blogspot.com/2022/06/cdc-and-ny-times-stopped-revealing.html.

83   "CDC Covid Data Tracker: Vaccine Effectiveness & Breakthrough
     Surveillance," Centers for Disease Control and Prevention, https://covid.cdc
     .gov/covid-data-tracker/#vaccine-effectiveness-breakthrough.

84   "Coronavirus in the U.S.: Latest Map and Case Count," *New York Times*,
     updated Jun. 6, 2022, https://www.nytimes.com/interactive/2021/us/covid
     -cases.html.

85   Guy Hatchard, PhD, "NZ Ministry of Health Data Shows Triple Vaccinated
     are Now More Vulnerable to Covid Infection and Hospitalisation
     than the Unvaccinated," *The Exposé*, Apr. 6, 2022, https://dailyexpose
     .uk/2022/04/06/nz-moh-data-triple-vaccinated-most-vulnerable-covid.

86   "Corona Virus in Israel- Key Metrics," *Israel Ministry of Health*, Mar. 2022,
     https://datadashboard.health.gov.il/COVID-19/general?utm_source=go.gov
     .il&utm_medium=referral. *Israeli COVID-19 Data Dashboard shows
     the Vast Majority of New COVID-19 cases among the vaccinated and
     vaccinated without validity groups. **Chrome will translate pages to
     English.

87   Heba N. Altarawneh, MD, Hiam Chermaitelly, PhD, Hussein H. Ayoub,
     PhD, et al., "Effects of Previous Infection and Vaccination on Symptomatic
     Omicron Infections," *NEJM*, (2022), https://www.doi.org/10.1056
     /NEJMoa2203965.

88   Marina Zhang, "Vaccination Increases Risk of COVID-19 Infection, but
     Infection without Vaccination Gives Immunity: Study," *The Epoch Times*,
     updated Jun. 22, 2022, https://www.theepochtimes.com/vaccination
     -increases-risk-of-covid-19-but-infection-without-vaccination-gives
     -immunity-study_4544042.html.

89   Katherine J. Reynolds, Corinna Pade, Joseph H. Gibbons, et al., "Immune
     Boosting by B.1.1.529 (Omicron) Depends on Previous SARS-CoV-2
     Exposure," *Science* (2022), https://www.doi.org/10.1126/science.abq1841.

90   Alex Berenson, "Don't read this if you're vaccinated," Substack, Jun. 16,
     2022, https://alexberenson.substack.com/p/dont-read-this-if-youre-
     vaccinated.

91   "'Original Antigenic Sin' Is a Real Problem with COVID-19 Vaccines,"
     Jeremy R. Hammond, Jun. 22, 2022, https://www.jeremyrhammond
     .com/2022/06/22/original-antigenic-sin-is-a-real-problem-with-covid-19
     -vaccines.

92   "Vaccinated 16 Times Less Likely to Die from Covid, Study Shows," *Bloomberg*,
     Nov. 8, 2021, https://www.bloomberg.com/news/articles/2021-11-09
     /vaccination-slashes-covid-deaths-and-icu-stays-16-fold-in-study.

93   Jake Tapper, "Sources: Biden Admin to Advise Booster Shots for Most U.S.
      Adults – Interview with Sanjay Gupta, *The Lead with Jake Tapper*," CNN,
      Aug. 17, 2022, 17:00 eastern, at the 17:35:12 mark in the show, https://
      transcripts.cnn.com/show/cg/date/2021-08-17/segment/02.

94   Pnina Shitrit, "Nosocomial Outbreak Caused by the Sars-Cov-2
      Delta Variant in a Highly Vaccinated Population, Israel, July 2021,"
      Eurosurveillance 26, no. 39 (2021), https://doi.org/10.2807/1560-7917.
      ES.2021.26.39.2100822.

95   Meredith Wadman, "A Grim Warning from Israel: Vaccination Blunts, but
      Does Not Defeat Delta," *Science*, Aug. 16, 2021, 6:55 PM, https://www
      .science.org/content/article/grim-warning-israel-vaccination-blunts-does
      -not-defeat-delt.

96   "Covid-19 Vaccine Surveillance Report–Week 8," *UK Health Security
      Agency*, Feb. 24, 2022, https://assets.publishing.service.gov.uk/government
      /uploads/system/uploads/attachment_data/file/1057599/Vaccine_surveillance
      report-_week-8.pdf.

97   "Whilst You've Been Distracted by Russia's Invasion, the UK Gov. Released
      a Report Confirming the Fully Vaccinated Now Account for 9 in Every
      10 Covid-19 Deaths in England," *The Exposé*, Mar. 1, 2022, https://
      dailyexpose.uk/2022/03/01/russia-distraction-uk-gov-revealed-triple
      -vaccinated-account-9-in-10-covid-deaths.

98   "COVID-19 Vaccination Statistics," NHS, Feb. 20, 2022, https://www
      .england.nhs.uk/statistics/wp-content/uploads/sites/2/2022/02/COVID-19
      -weekly-announced-vaccinations-24-February-2022.pdf.

99   Megan Munro, Charlotte Bermingham, Vahé Nafilyan, et al., "Deaths Involving
      COVID-19 by Vaccination Status, England: Deaths Occurring Between 1
      January 2021 and 31 January 2022," *Office of National Statistics*, Mar. 16,
      2022, https://www.ons.gov.uk/peoplepopulationandcommunity/births
      deathsandmarriages/deaths/bulletins/deathsinvolvingcovid19byvaccination
      statusengland/deathsoccurringbetween1january2021and31january2022.

100   "Covid-19 Weekly Surveillance in NSW: Epidemiological Week 03, Ending
      22 January 2022," *NSW Health*, Feb. 4, 2022, https://www.health.nsw.gov
      .au/Infectious/covid-19/Documents/covid-19-surveillance-report-20220204
      .pdf.

101   "Covid-19 Deaths Are at Record Levels in Australia and 4 in Every 5 of
      Them Are among the Fully Vaccinated," *The Exposé*, Feb. 15, 2022, https://
      dailyexpose.uk/2022/02/15/australia-4-in-5-covid-deaths-fully-vaccinated.

102   Josh Nicklaus and Nick Evershed, "Covid-19 Vaccine Australia Rollout
      Tracker: Percentage of Population Vaccinated and Vaccination Rate by
      State," *The Guardian*, Feb. 21, 2022, https://www.theguardian.com/australia
      -news/datablog/ng-interactive/2022/feb/21/covid-19-vaccine-rollout-australia

-vaccination-rate-progress-how-many-people-vaccinated-percent-tracker
-australian-states-number-total-daily-live-data-stats-updates-news-schedule
-tracking-chart-percentage-new-cases-today.

103   "Case Numbers, Spread and Deaths," *Government of Ontario*, last updated May 2, 2022, https://covid-19.ontario.ca/data/case-numbers-and-spread.

104   Oxford Martin School, "Coronavirus (COVID-19) Vaccinations-Haiti," *Our World in Data*, May 4, 2022, https://ourworldindata.org/covid-vaccinations ?country=HTI.

105   "COVID-19 Tracker-Haiti," Reuters, May 4, 2022, https://graphics.reuters .com/world-coronavirus-tracker-and-maps/countries-and-territories/haiti/.

106   Oxford Martin School, "Coronavirus (COVID-19) Vaccinations-Cuba," *Our World in Data*, May 4, 2022, https://ourworldindata.org/covid-vaccinations ?country=CUB.

107   "COVID-19 Tracker-Cuba," Reuters, May 17, 2022, https://graphics.reuters .com/world-coronavirus-tracker-and-maps/countries-and-territories/cuba/.

108   Mary Kekatos, "Who is Dying of COVID amid Omicron Surge and Widespread Vaccine Availability?" *ABC News*, Feb. 21, 2022, 4:05 AM, https://abcnews.go.com/Health/dying-covid-unvaccinated/story?id =82834971.

109   Christine S. Benn, DMSc, Frederik Schaltz-Buchholzer, MD, PhD, Sebastian Nielsen, MSc, et al., "Randomised Clinical Trials of COVID-19 Vaccines: Do Adenovirus-Vector Vaccines Have Beneficial Non-Specific Effects?" *The Lancet* (Preprint, 2022), https://papers.ssrn.com/sol3/papers .cfm?abstract_id=4072489.

110   Joseph Fraiman, Juan Erviti, Mark Jones, Sander Greenland, Patrick Whelan, Robert M. Kaplan, and Peter Doshi, "Serious Adverse Events of Special Interest Following mRNA Vaccination in Randomized Trials," Available at SSRN: https://ssrn.com/abstract=4125239.

111   "From the 7/1/2022 Release of VAERS Data: Found 1,329,135 Cases Where Vaccine Is COVID19," *MedAlerts/VAERS,* https://bit.ly/3yWVv19.

112   "From the 7/1/2022 Release of VAERS Data: Found 241,910 Cases Where Vaccine Is COVID19 and Serious," *MedAlerts/VAERS*, https://bit.ly /3PzcRY5.

113   Gloria Shwe Zin Tun, Dermot Gleeson, et al., "Immune-Mediated Hepatitis with the Moderna Vaccine, No Longer a Coincidence but Confirmed," *Journal of Hepatology* 76, no. 3 (2022): 738–753, www.doi.org/10.1016/j .jhep.2021.09.031.

114   Tobias Boettler, Benedikt Csernalabics, et al., "SARS-CoV-2 Vaccination Can Elicit a CD8 T-cell Dominant Hepatitis.," *Journal of Hepatology*, 2022 Apr 21:S0168-8278(22)00234-3, www.doi.org/10.1016/j.jhep .2022.03.040.

115   Celeste McGovern, "Medical Journals, VAERS Reports Show Girls as Young
        as 5 Developed Genital Ulcers after Pfizer Shot," *LifeSiteNews*, Apr. 8, 2022,
        https://www.lifesitenews.com/news/girls-as-young-as-5-years-old-developed
        -genital-ulcers-after-pfizer-shots.

116   "From the 7/1/2022 Release of VAERS Data: Found 2,084 Cases Where
        Vaccine Is COVID19 and Symptom Is Ischaemic Stroke," *MedAlerts/VAERS*,
        https://bit.ly/3yTVbAv.

117   "From the 7/1/2022 Release of VAERS Data: Found 20 Cases Where
        Vaccine Is COVID19 and Symptom Is Haemorrhagic Stroke," *MedAlerts
        /VAERS*, https://bit.ly/3wJuzAP.

118   "From the 7/1/2022 Release of VAERS Data: Found 15,120 Cases Where
        Vaccine Is COVID19 and Symptom Is Myocarditis," *MedAlerts/VAERS*,
        https://bit.ly/3LHFTla.

119   "From the 7/1/2022 Release of VAERS Data: Found 9,936 Cases Where
        Vaccine Is COVID19 and Symptom Is Seizures," *MedAlerts
        /VAERS*, https://bit.ly/3wHMo2S.

120   "From the 7/1/2022 Release of VAERS Data: Found 2,154 Cases Where
        Vaccine Is COVID19 and Symptom Is Paralysis," *MedAlerts
        /VAERS*, https://bit.ly/3wHhO8u.

121   "From the 7/1/2022 Release of VAERS Data: Found 282 Cases Where
        Vaccine Is COVID19 and Symptom Is Demyelination," *MedAlerts
        /VAERS*, https://bit.ly/3PIFmCI.

122   "From the 7/1/2022 Release of VAERS Data: Found 2,770 Cases Where
        Vaccine Is COVID19 and Symptom is Guillain-Barré Syndrome,"
        MedAlerts/VAERS, https://bit.ly/3wHipae.

123   "From the 7/1/2022 Release of VAERS Data: Found 6,533 Cases Where
        Vaccine Is COVID19 and Symptom is Bell's Palsy," *MedAlerts
        /VAERS*, https://bit.ly/3sVxdRA.

124   "From the 7/1/2022 Release of VAERS Data: Found 529 Cases Where
        Vaccine Is COVID19 and Symptom Is Herpes Simplex," *MedAlerts
        /VAERS*, https://bit.ly/3NxMbFk.

125   "From the 7/1/2022 Release of VAERS Data: Found 20 Cases Where
        Vaccine Is COVID19 and Symptom is Vaginal Lesions," *MedAlerts
        /VAERS*, https://bit.ly/3sUuFTC.

126   "From the 7/1/2022 Release of VAERS Data: Found 817 Cases Where
        Vaccine Is COVID19 and Symptom is Diabetes," *MedAlerts/VAERS*, https://
        bit.ly/39QsQQS.

127   "From the 7/1/2022 Release of VAERS Data: Found 4,521 Cases Where
        Vaccine Is COVID19 and Symptom Is Myocardial Infarction," *MedAlerts/
        VAERS*, https://bit.ly/38etirQ.

128 "From the 7/1/2022 Release of VAERS Data: Found 416 Cases Where Vaccine Is COVID19 and Symptom Is Hepatitis," *MedAlerts/VAERS*, bit. ly/3yR9Qwi.

129 Jessica Rose, PhD, MSc, BSc, "Critical Appraisal of VAERS Pharmacovigilance: Is the U.S. Vaccine Adverse Events Reporting System (VAERS) a Functioning Pharmacovigilance System?" *Science, Public Health Policy, and the Law* 3, (2021): 100-129, https://cf5e727d-d02d-4d71-89ff-9fe2d3ad957f.filesusr .com/ugd/adf864_0490c898f7514df4b6fbc5935da07322.pdf.

130 Victoria Furer, Devy Zisman, Adi Kibari, et al., "Herpes zoster following BNT162b2 mRNA COVID-19 vaccination in patients with autoimmune inflammatory rheumatic diseases: a case series," *Rheumatology* 60, no. SI (2021): S190-S195, https://doi.org/10.1093/rheumatology/keab345.

131 H. -H. Chiu, K. -C. Wei, A. Chen, and W. -H. Wang, "Herpes zoster following COVID-19 vaccine: a report of three cases," *QJM* 114, no. 7 (2021): 531-532, https://doi.org/10.1093/qjmed/hcab208.

132 Meiling Lee, "COVID-19 Vaccination Reactivates Highly Contagious Virus: Studies," *The Epoch Times*, Jun. 22, 2022, https://www.theepochtimes.com /reactivation-of-chickenpox-virus-following-covid-19-injections-on-the -rise_4549574.html.

133 Julia Stowe, et al., "Effectiveness of COVID-19 Vaccines Against Omicron and Delta Hospitalisation: Test Negative Case-Control Study," *medRxiv* (2022), https://doi.org/10.1101/2022.04.01.22273281.

134 Amanuensis, "Vaccinated Hospitalised for Non-Covid Reasons at FIVE Times the Rate of the Unvaccinated, U.K. Government Data Show," *The Daily Sceptic*, May 7, 2022, https://dailysceptic.org/2022/05/07 /vaccinated-hospitalised-for-non-covid-reasons-at-five-times-the-rate-of-the -unvaccinated-u-k-government-data-show.

135 Matthias Toying and Jana Olsen, "Charité Researcher Calls for Ambulances for Vaccine Victims," *MDR*, May 3, 2022, https://www.mdr.de/nachrichten /deutschland/panorama/corona-impfung-nebenwirkungen-impfschaeden -100.html.

136 "From the 7/1/2022 Release of VAERS Data: Found 29,273 Cases Where Vaccine Is COVID19 and Patient Died," *MedAlerts/VAERS*, https://bit.ly /3MMKVOq.

137 Ross Lazarus et al., "Electronic Support for Public Health–Vaccine Adverse Event Reporting System (ESP:VAERS)," *The Agency for Healthcare Research and Quality (AHRQ)* (2010), https://digital.ahrq.gov/sites/default/files/docs /publication/r18hs017045-lazarus-final-report-2011.pdf.

138 Tom Shimabukuro, Michael Nguyen, David Martin, and Frank DeStefano, "Safety Monitoring in the Vaccine Adverse Event Reporting System (VAERS),"

Vaccine 33, no. 36 (2015): 4398–4405, https://doi.org/10.1016/j.vaccine
.2015.07.035.

139   Nolan E. Bowman, "German Insurance Company Fires CEO Who Released
       COVID Vaccine Injury Data, Then Scrubs Data From Website," *The
       Defender*, Mar. 14, 2022, https://childrenshealthdefense.org/defender
       /german-insurance-fires-andreas-schofbeck-covid-vaccine-injuries-data.

140   Steve Kirsch, "License to Kill (and How to Redpill Patients)," *Substack*, Mar.
       5, 2022, https://stevekirsch.substack.com/p/license-to-kill.

141   Steve Kirsch, "Latest VAERS Estimate: 388,000 Americans Killed by the
       COVID Vaccines," *Substack*, Dec. 14, 2021, https://stevekirsch.substack.com
       /p/latest-vaers-estimate-388000-americans.

142   Mark Skidmore, "How Many People Died from the Covid-19
       Inoculations?" Feb. 28, 2022, https://mark-skidmore.com/wp-content
       /uploads/2022/02/Survey-of-Covid-Heatlh-Experiences-Final.pdf.

143   "AMERICA'S FRONTLINE DOCTORS vs. XAVIER BECERRA,"
       Secretary of the U.S. Department of Health and Human Services, Civil
       Action No. 2:21-cv-00702-CLM (2021), https://fossaorg.files.wordpress.com
       /2021/07/m-for-pi-file-stamped.pdf.

144   S. J. Thomas et al., "Supplementary Appendix: Safety and Efficacy of the
       BNT162b2 mRNA Covid-19 Vaccine through 6 months," *NEJM* 385,
       (2021): 1761-1773, S4 Table, https://www.nejm.org/doi/suppl/10.1056
       /NEJMoa2110345/suppl_file/nejmoa2110345_appendix.pdf.

145   Peter Doshi, "Peter Doshi: Pfizer and Moderna's '95% Effective' Vaccines—
       Let's Be Cautious and First See the Full Data," *The BMJ Opinion*, Nov. 26,
       2020, https://blogs.bmj.com/bmj/2020/11/26/peter-doshi-pfizer-and
       -modernas-95-effective-vaccines-lets-be-cautious-and-first-see-the-full-data.

146   Matthew Herper, "Pfizer and BioNTech Speed Up Timeline for Offering
       Covid-19 Vaccine to Placebo Volunteers," *STAT*, Jan. 1, 2021, https://www
       .statnews.com/2021/01/01/pfizer-and-biontech-speed-up-timeline-for
       -offering-covid-19-to-placebo-volunteers.

147   Abbie Boudreau and Scott Zamost, "Ex-CDC Head Recalls '76 Swine Flu
       Outbreak," *CNN*, https://edition.cnn.com/2009/HEALTH/04/30/swine
       .flu.1976/index.html.

148   Justus R. Hope, MD, and Robert Malone, MD, "Life Insurance Deaths Up
       40%—Dr. Robert Malone's Chilling Analysis," *The Desert Review*, Jan. 6,
       2022, https://www.thedesertreview.com/opinion/columnists/life-insurance
       -deaths-up-40---dr-robert-malone-s-chilling-analysis/article_d24bccac-6f38
       -11ec-912f-1f6d8fc5fac4.html.

149   Melody Schreiber, "True Number of Covid Deaths in the US Probably
       Undercounted, Experts Say," *The Guardian*, Jan. 7, 2022, https://www

.theguardian.com/us-news/2022/jan/07/true-number-covid-deaths-us
-likely-undercounted-experts.

150   Margaret Menge, "BREAKING: Fifth Largest Life Insurance Company in
       the US Paid Out 163% More for Deaths of Working People Ages 18–64
       in 2021 - Total claims/benefits up $6 BILLION," *Substack*, Jun. 15, 2022,
       https://crossroadsreport.substack.com/p/breaking-fifth-largest-life-insurance.

151   Avneet Kaur, "Life Insurance Death Claims Shoot 41%, Up 3.5x in 2021,"
       Fortune India, Dec. 30, 2021, https://www.fortuneindia.com/enterprise/life
       -insurance-death-claims-shoot-41-up-35x-in-2021/106563.

152   RFK Jr. and Edward Dowd, "The Defender Show Episode 33-Pfizer Fraud
       and Wall Street with Ed Dowd," *CHD-TV*, Mar. 22, 2022, 00:12:46–
       00:13:03, https://live.childrenshealthdefense.org/shows/the-defender-show
       /DthWZW2-ZJ.

153   Seth Hancock, "40% Rise Nationwide in Excess Deaths among 18-to-
       49-Year-Olds, CDC Data Show," *The Defender*, Jan. 20, 2022, https://
       childrenshealthdefense.org/defender/rise-nationwide-excess-deaths-18-to
       -49-year-olds.

154   "1100% Increase in U.S. Military Deaths… U.S. Lawyer Todd Callender
       Speaks Out," BitChute, Mar. 21, 2022, https://www.bitchute.com/video
       /b6s8lrRi2T5v.

155   Disabled Rights Advocates, "Todd Callender," (2022), https://dradvocates
       .com/lawyers.

156   David Horowitz, "Horowitz: More VAERS-Reported Vaccine Deaths in
       Our Military than COVID Deaths," *Blaze Media*, Op-Ed, Mar. 21, 2022,
       https://www.theblaze.com/op-ed/horowitz-more-vaers-reported-vaccine
       -deaths-in-our-military-than-covid-deaths.

157   "COVID-19 Weekly Total Deaths Archive" NHS, COVID-19 total
       announced deaths August 12, 2021 and April 7, 2022 weekly data files,
       https://web.archive.org/web/20220629190134/https://www.england.nhs
       .uk/statistics/statistical-work-areas/covid-19-daily-deaths/weekly-total
       -archive.

158   Margaret Menge, "Indiana Life Insurance CEO Says Deaths Are Up 40%
       Among People Ages 18-64," *The Center Square*, Jan. 1, 2022, https://www
       .thecentersquare.com/indiana/indiana-life-insurance-ceo-says-deaths-are
       -up-40-among-people-ages-18-64/article_71473b12-6b1e-11ec-8641
       -5b2c06725e2c.html.

159   "992 Athlete Cardiac Arrests, Serious Issues, 644 Dead, after COVID Shot,"
       Real Science, Apr. 26, 2022, https://goodsciencing.com/covid/athletes-suffer
       -cardiac-arrest-die-after-covid-shot.

160   "Prevent Sudden Cardiac Arrest in Student Athletes," *California Interscholastic
       Federation*, Aug. 21, 2021, https://cifstate.org/sports-medicine/sca/SCA_Slideline
       _Action_Plan_CIF.pdf.

161 *Frontline* News Staff, "Israel Education Ministry Ordered 1,556
     Defibrillators Installed in All Schools with over 500 Children in Preparation
     for Children's Vaccine Campaign," *Frontline News*, Feb. 14, 2022, https://
     americasfrontlinedoctors.org/news/post/israel-education-ministry-ordered
     -1556-defibrillators-installed-in-all-schools-with-over-500-children-in
     -preparation-for-childrens-vaccine-campaign.

162 Orange County Public Schools (OCPS), "Sports Physicals," *OCPS* (2022),
     https://www.ocps.net/departments/Athletics/sports_physicals.

163 Joseph Mercola, "The Latest Tragedy: Sudden Adult Death Syndrome," *The
     Epoch Times*, Jun. 20, 2022, https://www.theepochtimes.com/the-latest
     -tragedy-sudden-adult-death-syndrome_4545014.html.

164 Tom Heaton, "Healthy Young People are Dying Suddenly and Unexpectedly
     from a Mysterious Syndrome - as Doctors Seek Answers through a New
     National Register," *Daily Mail*, Jun. 8, 2022, https://www.dailymail.co.uk
     /news/article-10895067/Doctors-trying-determine-young-people-suddenly
     -dying.html.

165 Christopher L. F. Sun, Eli Jaffe, and Letsef Levi, "Increased Emergency
     Cardiovascular Events among Under-40 Population in Israel during Vaccine
     Rollout and Third COVID-19 Wave," *Nature* 12, no. 9678, (2022), https:
     //www.nature.com/articles/s41598-022-10928-z.

166 Y. Rabinovitz, "New study links COVID vaccines to 25% increase in cardiac
     arrest for both males & females," 7 *Israel National News*, Jun. 1, 2022,
     https://www.israelnationalnews.com/news/328529.

167 Øystein Karlstad, MScPharm, PhD, Petteri Hovi, MD, PhD, et al., "SARS-
     CoV-2 Vaccination and Myocarditis in a Nordic Cohort Study of 23 Million
     Residents," *JAMA* (2022), https://jamanetwork.com/journals
     /jamacardiology/article-abstract/2791253.

168 Michael Kang and Jason An, "Virus Myocarditis," *StatPearls* [Internet],
     Treasure Island (FL): StatPearls Publishing, Jan. 2022, PMID: 29083732,
     https://www.ncbi.nlm.nih.gov/books/NBK459259.

169 Steve Kirsch, "New Study from Germany Confirms Higher Vax Coverage
     –Higher Excess Mortality," *Substack*, Nov. 19, 2021, https://stevekirsch
     .substack.com/p/new-study-from-germany-confirms-higher.

170 John Gibson, "The Rollout of COVID-19 Booster Vaccines is Associated
     With Rising Excess Mortality in New Zealand," *University of Waikato*, Jun.
     2022, 7, https://repec.its.waikato.ac.nz/wai/econwp/2211.pdf.

171 Apoorva Mandavilli, "The C.D.C. Isn't Publishing Large Portions of the
     Covid Data It Collects," *New York Times*, Feb. 20, 2022, https://www.nytimes
     .com/2022/02/20/health/covid-cdc-data.html.

172 Michael Nevradakis, PhD, "FDA Now Wants 75 Years to Release
     Pfizer Vaccine Documents," *The Defender*, Dec. 10, 2021, https://

childrenshealthdefense.org/defender/fda-75-years-release-pfizer-vaccine
-documents.

173 Ben Fidler, "Top FDA Vaccine Officials to Leave Agency as Decision on
COVID-19 Booster Looms," *BioPharma Dive*, Aug. 31, 2021, https://www
.biopharmadive.com/news/marion-gruber-phil-krause-leaving-fda/605859.

174 "Trusted News Initiative (TNI) to Combat Spread of Harmful Vaccine
Disinformation and Announces Major Research Project," *BBC*, Dec. 10,
2020, https://www.bbc.com/mediacentre/2020/trusted-news-initiative
-vaccine-disinformation.

175 Beth Snyder Bulik, "The Top 10 Ad Spenders in Big Pharma for 2020,"
Fierce Pharma, Apr. 19, 2021, https://www.fiercepharma.com/special-report
/top-10-ad-spenders-big-pharma-for-2020#:~:text=Pharma%20TV%20
advertising%20remained%20the%20cornerstone%20of%20spending%20
with%20%244.58,was%2073%25%20of%20pharma's%20investment.

176 Alan McLeod, "Documents Show Bill Gates Has Given $319 Million to
Media Outlets to Promote His Global Agenda," *The Gray Zone*, Nov. 21,
2021, https://thegrayzone.com/2021/11/21/bill-gates-million-media
-outlets-global-agenda.

177 Chris Pandolfo, "Exclusive: The Federal Government Paid Hundreds of
Media Companies to Advertise the COVID-19 Vaccines While Those Same
Outlets Provided Positive Coverage of the Vaccines," *The Blaze*, Mar. 3,
2022, https://www.theblaze.com/news/review-the-federal-government-paid
-media-companies-to-advertise-for-the-vaccines.

178 Marty Makary, "The CDC — Which Is Withholding Information — Has a
Hidden Agenda," *New York Post*, Feb. 27, 2022, https://nypost.com/2022
/02/27/the-cdc-has-a-hidden-agenda-when-it-comes-to-covid-vaccines.

179 Marty Makary, MD, *The Price We Pay: What Broke American Health Care
and How to Fix It* (Bloomsbury Publishing, Sep. 10, 2019), https://www
.amazon.com/Price-We-Pay-American-Care/dp/1635574110.

180 Jeremy R. Hammond, "A Chronicle of Statements about Natural Immunity
Once Deemed Misinformation But Now Admittedly True," *Jeremy R.
Hammond*, Feb. 21, 2022, https://jeremyrhammond.com/2022/02/21/a
-chronicle-of-statements-about-natural-immunity-once-deemed
-misinformation-but-now-admittedly-true.

181 Paul Alexander, "More than 150 Comparative Studies and Articles on Mask
Ineffectiveness and Harms," *Brownstone Institute*, Dec. 20, 2021, https://
brownstone.org/articles/more-than-150-comparative-studies-and-articles
-on-mask-ineffectiveness-and-harms.

182 Daniel Politi, "CDC Updates Guidelines to Make Clear Cloth Masks Offer
Least Protection against COVID," *Slate*, Jan. 15, 2022, https://slate.com
/news-and-politics/2022/01/cdc-guidelines-cloth-masks-n95-kn95-covid.html.

183 Beny Spira, "Correlation Between Mask Compliance and COVID-19 Outcomes in Europe," *Cureus* 14, no. 4, (April 19, 2022): e24268, www.doi.org/10.7759/cureus.24268.

184 Alicia Powe, "Another Study Confirms Wearing Masks Increases COVID Infection," *Gateway Pundit*, May 20, 2022, https://www.thegatewaypundit.com/2022/05/another-study-confirms-wearing-masks-increases-infection-covid.

185 Lauren C. Jenner, Jeanette M. Rotchell, Robert T. Bennett, et al., "Detection of Microplastics in Human Lung Tissue Using μFTIR Spectroscopy," *Science of the Total Environment* 831 (2022), https://www.sciencedirect.com/science/article/pii/S0048969722020009?via%3Dihub.

186 Ambarish Chandra, Tracy Beth Høeg, "Revisiting Pediatric COVID-19 Cases in Counties With and Without School Mask Requirements—United States, July 1—October 20, 2021," Available at SSRN: https://ssrn.com/abstract=4118566 or http://dx.doi.org/10.2139/ssrn.4118566.

187 S. E. Budzyn, M. J. Panaggio, S. E. Parks, et al., "Pediatric COVID-19 Cases in Counties With and Without School Mask Requirements—United States, July 1–September 4, 2021," MMWR Morb Mortal Wkly Rep 2021;70:1377–1378. DOI: http://dx.doi.org/10.15585/mmwr.mm7039e3external icon.

188 David Leonhardt, "Why Masks Work, but Mandates Haven't," *New York Times*, May 31, 2022, https://www.nytimes.com/2022/05/31/briefing/masks-mandates-us-covid.html.

189 Jon Miltimore, "The CDC Changed Its COVID Risk Formula. The Results Are Stunning." *FEE Stories*, Mar. 3, 2022, https://fee.org/articles/the-cdc-changed-its-covid-risk-formula-the-results-are-stunning.

190 Graison Dangor, "CDC's Six-Foot Social Distancing Was 'Arbitrary,' Says Former FDA Commissioner," *Forbes*, Sep. 19, 2021, https://www.forbes.com/sites/graisondangor/2021/09/19/cdcs-six-foot-social-distancing-rule-was-arbitrary-says-former-fda-commissioner/?sh=5fe26e0de8e6.

191 Yael Halon, "Former CDC Director Redfield: The Safest Place for Children Right Now Is in the Classroom," *FOX News*, Jan. 5, 2022, https://www.foxnews.com/media/former-cdc-director-redfield-safest-place-children-classroom.

192 Dr. Megan Kuhfeld and Dr. Beth Tarasawa, "The COVID-19 Slide: What Summer Learning Loss Can Tell Us about the Potential Impact of School Closures on Student Academic Achievement," *NWEA Research*, Apr. 2020, 2, https://www.nwea.org/content/uploads/2020/05/Collaborative-Brief_Covid19-Slide-APR20.pdf.

193 Giorgio D. Pietro, Federico Biagi, Patricia Dinis Mota Da Costa, et al., "JRC Technical Report: The Likely Impact of COVID-19 on Education: Reflections Based on the Existing Literature and Recent International

Datasets," *European Commission*, 2020, https://publications.jrc.ec.europa.eu
/repository/handle/JRC121071.

194   Dimitri A. Christakis, MD, MPH, Wil Van Cleve, MD, MPH, and
Frederick J. Zimmerman, PhD, "Estimation of US Children's Educational
Attainment and Years of Life Lost Associated With Primary School Closures
During the Coronavirus Disease 2019 Pandemic," *JAMA* 3, no. 11 (2020):
e2028786, https://www.doi.org/10.1001/jamanetworkopen.2020.28786.

195   "UN Secretary-General Warns of Education Catastrophe, Pointing to
UNESCO Estimate of 24 Million Learners at Risk of Dropping Out,"
UNESCO, Jun. 8, 2020, https://en.unesco.org/news/secretary-general
-warns-education-catastrophe-pointing-unesco-estimate-24-million
-learners-0.

196   UNICEF, "COVID:19 Scale of Education Loss 'nearly insurmountable',
Warns UNICEF," media factsheet, Jan. 23, 2022, https://www.unicef.org
/press-releases/covid19-scale-education-loss-nearly-insurmountable-warns
-unicef.

197   Carl Henneghan, Jon Brassey, and Tom Jefferson, "CG REPORT 3: The
Impact of Pandemic Restrictions on Childhood Mental Health," *Collateral
Global*, Oct. 2, 2021, https://collateralglobal.org/article/report-the-impact
-of-pandemic-restrictions-on-childhood-mental-health.

198   "The Ongoing Impact of the Pandemic on Children," *Health Advisory &
Recovery Team*, Oct. 11, 2021, https://www.hartgroup.org/the-ongoing
-impact-of-the-pandemic-on-children.

199   "Mental Health of Children and Young People in England, 2020: Wave 1
Follow up to the 2017 Survey," NHS Digital, Oct. 22, 2020, https://digital.
nhs.uk/data-and-information/publications/statistical/mental-health-of
-children-and-young-people-in-england/2020-wave-1-follow-up.

200   "Young People 'unable to cope with life' Since Pandemic, Warns Prince's
Trust," *Prince's Trust*, Jan. 19, 2021, https://www.princes-trust.org.uk/about
-the-trust/news-views/tesco-youth-index-2021.

201   "The Ongoing Impact of the Pandemic on Children," *Health Advisory &
Recovery Team*, Oct. 11, 2021, https://www.hartgroup.org/the-ongoing
-impact-of-the-pandemic-on-children.

202   Camila Turner, "'Explosion' of Children with Tics and Tourette's from
Lockdown," *The Telegraph*, Feb. 13, 2021, https://www.telegraph.co.uk
/news/2021/02/13/explosion-children-tics-tourettes-lockdown.

203   Kat Ley, "Record Numbers of Children are Being Prescribed
Antidepressants," *The Times*, Aug. 31, 2022, https://www.thetimes.co.uk
/article/record-numbers-of-children-are-being-prescribed-antidepressants
-88268m0rt.

204   "The State of the Global Education Crisis: A Path To Recovery," *UNESCO, UNICEF, World Bank*, joint report, 2021, https://www.unicef.org/media /111621/file/%20The%20State%20of%20the%20Global%20Education %20Crisis.pdf%20.pdf.

205   Hugh McCarthy, "Lockdowns Killed Hundreds of Thousands of Children, Says the UN–Was it Really Worth It?" *Sign of the Times*, Jun. 15, 2022, https://www.sott.net/article/468820-Lockdowns-killed-hundreds-of -thousands-of-children-says-the-UN-was-it-really-worth-it.

206   Lizzy Davies, "'Lost Generation': Education in Quarter of Countries at Risk of Collapse, Study Warns," *The Guardian*, Sep. 6, 2021, https://www.the guardian.com/global-development/2021/sep/06/lost-generation-education -in-quarter-of-countries-at-risk-of-collapse-study-warns.

207   "Coronavirus Disease (COVID-19): Herd Immunity, Lockdowns and COVID-19," *World Health Organization (WHO)*, Dec. 31, 2020, https: //www.who.int/news-room/questions-and-answers/item/herd-immunity -lockdowns-and-covid-19.

208   FOX News, "A Doctor a Day Letter - Signed," *Scribd*, May 19, 2020, https://www.scribd.com/document/462319362/A-Doctor-a-Day-Letter -Signed.

209   Martin Kulldorff, PhD, Sunetra Gupta, PhD, Jay Bhattacharya, PhD, "The Great Barrington Declaration," Great Barrington Declaration, Oct. 4, 2020, https://gbdeclaration.org/#read.

210   Kristalina Georgieva, "Urgent Action Needed to Address a Worsening 'Two-Track' Recovery," *IMF Blog*, Jul. 27, 2021, https://blogs.imf.org/2021/07 /07/urgent-action-needed-to-address-a-worsening-two-track-recovery.

211   Jonas Herby, Lars Jonung, and Steve H. Hanke, "A Literature Review and Meta-Analysis of the Effects of Lockdowns on COVID-19 Mortality," *SAE Johns Hopkins Institute for Applied Economics, Global Health, and the Study of Business Enterprise*, Jan. 2022, https://sites.krieger.jhu.edu/iae/files/2022/01 /A-Literature-Review-and-Meta-Analysis-of-the-Effects-of-Lockdowns-on -COVID-19-Mortality.pdf.

212   James Grant, "Proof That Blue States Did Fail Their People During Pandemic: Harsh Lockdowns Caused Huge Deaths Rates, Ruined Kids' Education and Destroyed Business, Bombshell Research Finally Shows - With Ny, Nj, Ca and Il All Receiving An F-Grade," *The Daily Mail*, Apr. 11, 2022, https://www.dailymail.co.uk/news/article-10708225/The-states -FAILED-protect-people-COVID.html.

213   Phillip W. Magness and Peter C. Earle, "The Fickle 'Science' of Lockdowns," *Wall Street Journal*, Dec. 19, 2021, https://www.wsj.com/articles/lockdown -science-pandemic-imperial-college-london-quarantine-social-distance -covid-fauci-omicron-11639930605.

214 @disclosetv, "NEW - Fauci: 'You use lockdowns to get people vaccinated.'" *Twitter*, Apr. 14, 2022, https://twitter.com/disclosetv/status/1514670249316077569?s=20&t=a6k0PkHc13_I2XRHDvydxw.

215 ICAN, "CDC Cannot Provide an Instance of a Single Confirmed COVID-19 Death in a Child Younger Than 16," legal update, Mar. 29, 2022, https://www.icandecide.org/ican_press/cdc-cannot-provide-an-instance-of-a-single-confirmed-covid-19-death-in-a-child-younger-than-16.

216 AL Sorg et al., "Risk of Hospitalization, Severe Disease, and Mortality Due to COVID-19 and PIMS-TS in Children with SARS-CoV-2 Infection in Germany," *medRxiv*, Nov. 30, 2021, https://www.medrxiv.org/content/10.1101/2021.11.30.21267048v1.

217 Office for National Statistics, "COVID-19 Deaths and Autopsies Feb 2020 to Dec 2021," Jan. 17, 2022, https://www.ons.gov.uk/aboutus/transparencyandgovernance/freedomofinformationfoi/covid19deathsandautopsiesfeb2020todec2021.

218 Marty Makary, "Risk Factors for COVID-19 Mortality among Privately Insured Patients," *Fair Health Media*, Nov. 11, 2020, https://s3.amazonaws.com/media2.fairhealth.org/whitepaper/asset/Risk%20Factors%20for%20COVID-19%20Mortality%20among%20Privately%20Insured%20Patients%20-%20A%20Claims%20Data%20Analysis%20-%20A%20FAIR%20Health%20White%20Paper.pdf.

219 Alexander C. Dowell et al., "Children Develop Robust and Sustained Cross-Reactive Spike-Specific Immune Responses to SARS-CoV-2 Infection," *Nature*, Dec. 22, 2021, https://www.nature.com/articles/s41590-021-01089-8.

220 COVID-19 Forecasting Team, "Variation in the COVID-19 Infection–Fatality Ratio by Age, Time, and Geography during the Pre-Vaccine Era: A Systematic Analysis," *The Lancet*, Feb. 24, 2022, https://www.thelancet.com/journals/lancet/article/PIIS0140-6736(21)02867-1/fulltext.

221 Gilbert T Chua et al., "Epidemiology of Acute Myocarditis/Pericarditis in Hong Kong Adolescents Following Comirnaty Vaccination," *Clinical Infectious Diseases*, Nov. 28, 2021, https://academic.oup.com/cid/advance-article/doi/10.1093/cid/ciab989/6445179?login=false.

222 Katie A Sharff et al., "Risk of Myopericarditis Following COVID-19 mRNA Vaccination in a Large Integrated Health System: A Comparison of Completeness and Timeliness of Two Methods," *medRxiv*, Dec. 27, 2021, https://www.medrxiv.org/content/10.1101/2021.12.21.21268209v1.

223 "Comparison of Deaths of UK 10 to 17 Year-Olds Between Unvaccinated and Those Who Had Two Shots," Dr. Wayne Winston, PhD, professor emeritus of Decision Sciences at the University of Indiana's Kelley School of

Business, https://childrenshealthdefense.org/wp-content/uploads/Comparison
-of-Deaths-of-UK-10-bsh-rev.pdf.

224   Jenny Strasburg and Dominic Chopping, "Some European Countries Are
Limiting the Use of Moderna's COVID-19 Vaccine in Younger Ages," *Wall
Street Journal*, Oct. 7, 2021, https://www.wsj.com/articles/some-european
-countries-are-limiting-the-use-of-modernas-covid-19-vaccine-11633610069.

225   TrialSite Staff, "Mother of Maddie De Garay Speaks Out about Her 13 Year
Old Daughter's Life Altering Injuries From Pfizer's Covid Vaccine," *TrialSite
News*, Dec. 11, 2021, https://trialsitenews.com/mother-of-maddie-de-garay
-speaks-out-about-her-13-year-old-daughters-life-altering-injuries-from-pfizers
-covid-vaccine. *This document is behind a paywall, but you can read it
here, https://childrenshealthdefense.org/wp-content/uploads/Mother-of
-Maddie-de-Garay-speaks-out-about-her-13-year-old-daughters-life-altering
-injuries-from-Pfizers-Covid-vaccine..pdf.

226   P. D. Thacker, "Covid-19: Researcher Blows the Whistle on Data Integrity
Issues in Pfizer's Vaccine Trial," *The BMJ* 375, (2021): 2635, doi:10.1136/
bmj.n2635, https://www.bmj.com/content/375/bmj.n2635.

227   Rachel Maddow, "Interview with Anthony Fauci, *The Rachel Maddow
Show*," *MSNBC*, Dec. 29, 2021, https://www.msnbc.com/transcripts/
transcript-rachel-maddow-show-12-29-21-n1286909.

228   Glenn Braunstein, Lori Schwartz, et al., "False Positive Results With
SARS-CoV-2 RT-PCR Tests and How to Evaluate a RT-PCR-Positive Test
for the Possibility of a False Positive Result," *Journal of Occupational and
Environmental Medicine* 63, no 3 (2021): e159-e162, doi.org/10.1097/JOM
.0000000000002138. *PCR test can have a high rate of false positive results,
especially when COVID prevalence is low.

229   Eva Frederick, "New Research Reveals Why Some Patients May Test Positive
for Covid-19 Long After Recovery," *Whitehead Institute*, May 6, 2021,
https://wi.mit.edu/news/new-research-reveals-why-some-patients-may-test
-positive-covid-19-long-after-recovery. *Some people test positive on PCR
for months after an infection.

230   Health and Human Services, "HHS Is Releasing $9 Billion in Provider
Relief Fund Payments to Support Health Care Providers Affected by the
COVID-19 Pandemic," press release, Dec. 14, 2021, https://www.hhs.
gov/about/news/2021/12/14/hhs-releasing-9-billion-in-prf-payments-to-
support-providers-affected-by-covid-19.html.

231   Elizabeth Lee Vliet, M.D. and Ali Shultz, J.D., "Biden's Bounty on Your
Life: Hospitals' Incentive Payments for COVID-19," *Association of American
Physicians and Surgeons*, Nov. 17, 2021, https://aapsonline.org/bidens
-bounty-on-your-life-hospitals-incentive-payments-for-covid-19/.

232 Oversight Committee, "Select Subcommittee Hearing 'The Urgent Need for a National Plan to Contain the Coronavirus,'" YouTube, 02:36:11-02:37:56, Jul. 31, 2020, https://www.youtube.com/watch?v=YkP1t_2u5B0.

233 "Conditions Contributing to Deaths Involving COVID-19, by Age Group, United States. Week Ending 2/1/2020 to 12/5/2020," CDC, Dec. 6, 2020, https://www.cdc.gov/nchs/data/health_policy/covid19-comorbidity-expanded-12092020-508.pdf.

234 Steven Nelson, "COVID Origins Report Says It's 'plausible' Virus Leaked from Wuhan Lab," *New York Post*, Aug. 27, 2021, https://nypost.com/2021/08/27/covid-origins-report-says-its-plausible-virus-leaked-from-wuhan-lab.

235 Katherine Eban, "'This Shouldn't Happen': Inside the Virus-Hunting Nonprofit at the Center of the Lab-Leak Controversy," *Vanity Fair*, Mar. 31, 2022, https://www.vanityfair.com/news/2022/03/the-virus-hunting-nonprofit-at-the-center-of-the-lab-leak-controversy.

236 Jeremy R. Hammond, "The CDC Finally Admits That Natural Immunity to SARS-CoV-2 is Superior to the Immunity Induced by COVID-19 Vaccines," *Jeremy R. Hammond*, Feb. 10, 2022, https://www.jeremyrhammond.com/2022/02/10/the-cdc-finally-admits-that-natural-immunity-to-sars-cov-2-is-superior-to-the-immunity-induced-by-covid-19-vaccines.

237 Sivan Gazit et al., "Comparing SARS-CoV-2 Natural Immunity to Vaccine-Induced Immunity: Reinfections Versus Breakthrough Infections," *medRxiv*, Aug. 25, 2021, https://www.medrxiv.org/content/10.1101/2021.08.24.21262415v1.

238 Washington Journal, "Dr. Fauci: Influenza Vaccine," *C-SPAN*, 00:30:07, Oct. 11, 2004, https://www.c-span.org/video/?183885-2/influenza-vaccine.

239 "Myths and Facts about COVID-19 Vaccines," *CDC*, Updated Dec. 15, 2021, https://www.cdc.gov/coronavirus/2019-ncov/vaccines/facts.html.

240 Markus Alden et al., "Intracellular Reverse Transcription of Pfizer BioNTech COVID-19 mRNA Vaccine BNT162b2 In Vitro in Human Liver Cell Line," *Curr. Issues Mol. Bio* 44 (Feb. 25, 2022): 1115-26, https://doi.org/10.3390/cimb44030073.

241 Informed Consent Action Network, "CDC Cannot Back up Its 'Facts' about Genetic Manipulation from COVID-19 Vaccines," *ICAN Press Release*, Apr. 13, 2022, https://www.icandecide.org/ican_press/cdc-cannot-back-up-its-facts-regarding-potential-genetic-mutation-from-covid-19-vaccines.

242 Chase Peterson-Withorn, "Nearly 500 People Became Billionaires During the Pandemic Year," *Forbes*, Apr. 6, 2021, https://www.forbes.com/sites/chasewithorn/2021/04/06/nearly-500-people-have-become-billionaires-during-the-pandemic-year/?sh=7f03f49025c0.

243 Juliana Kaplan, "Billionaires Made $3.9 Trillion during the Pandemic — Enough to Pay for Everyone's Vaccine," *Insider*, Jan. 26, 2021, https://www.businessinsider.com/billionaires-made-39-trillion-during-the-pandemic-coronavirus-vaccines-2021-1.

244 OXFAM International, "Mega-rich Recoup COVID-Losses in Record-Time Yet Billions Will Live in Poverty for at Least a Decade," press release, Jan. 25, 2021, https://www.oxfam.org/en/press-releases/mega-rich-recoup-covid-losses-record-time-yet-billions-will-live-poverty-least.

245 Mohammed Yusef, "Africa: Global Pandemic Increased Poverty in Africa—Report," AllAfrica, May 16, 2022, https://allafrica.com/stories/202205170024.html.

246 David Bell, "The Corruption of the World Health Organization," The Brownstone Institute, May 27, 2022, https://brownstone.org/articles/the-corruption-of-the-world-health-organization.

247 R. Fairlie, "The Impact of COVID-19 on Small Business Owners: Evidence from the First 3 Months after Widespread Social-Distancing Restrictions," *J Econ Manag Strategy*, Aug. 27, 2020, https://doi.org/10.1111/jems.12400.

248 Sean CL Deoni et al., "Impact of the COVID-19 Pandemic on Early Child Cognitive Development: Initial Findings in a Longitudinal Observational Study of Child Health," *medRxiv*, Aug. 11, 2021, https://www.medrxiv.org/content/10.1101/2021.08.10.21261846v1.full.pdf.

249 Megan Redshaw, "8,817 COVID Vaccine Injuries Reported to CDC Among Kids 5 to 11, as Study Shows Pfizer Vaccine Only 12% Effective in That Age Group," *The Defender*, Mar. 4, 2022, https://childrenshealthdefense.org/defender/vaers-cdc-covid-vaccine-injuries-kids-5-to-11-pfizer-vaccine.

250 Clare Smith et al., "Deaths in Children and Young People in England after SARS-CoV-2 Infection during the First Pandemic Year," *Nature Medicine*, Jan. 2022, https://www.nature.com/articles/s41591-021-01578-1.pdf.

251 HLPE, "Impacts of COVID-19 on Food Security and Nutrition: Developing Effective Policy Responses to Address the Hunger and Malnutrition Pandemic," *CFS Committee on World Food Security*, Sep. 2020, https://www.fao.org/3/cb1000en/cb1000en.pdf.

252 *SkyNews*, "COVID-19: Eleven People Dying from Hunger a Minute as Pandemic Fuels Starvation Crisis, Warns Oxfam," *SkyNews*, Jul. 9, 2021, https://news.sky.com/story/covid-19-eleven-people-dying-from-hunger-a-minute-as-pandemic-fuels-starvation-crisis-warns-oxfam-12352071.

253 "COVID-19 and Children," UNICEF, 2021, https://data.unicef.org/covid-19-and-children.

254   Sophie Cousins, "2.5 Million More Child Marriages Due to COVID-19 Pandemic, The Lancet, Oct. 10, 2020, https://www.thelancet.com/journals /lancet/article/PIIS0140-6736(20)32112-7/fulltext.

255   Surya Kant and Richa Tyagi, "The Impact of COVID-19 on Tuberculosis: Challenges and Opportunities," Therapeutic Advances in Infectious Disease, Jun. 9, 2021, https://www.ncbi.nlm.nih.gov/labs/pmc/articles /PMC8193657.

256   "COVID-19 and HIV: A Tale of Two Pandemics," International AIDS Society, Jul. 2020, https://covid19andhivreport.iasociety.org.

257   WHO, "More Malaria Cases and Deaths in 2020 Linked to COVID-19 Disruptions," news release, Dec. 6, 2021, https://www.who.int/news /item/06-12-2021-more-malaria-cases-and-deaths-in-2020-linked-to-covid -19-disruptions.

258   "Direct and Indirect Effects of the COVID-19 Pandemic and Response in South Asia," UNICEF, Mar. 2021, https://www.unicef.org/rosa/media /13066/file/Main%20Report.pdf.

259   "COVID-19 Disruptions Killed 228,000 Children in South Asia, Says UN Report," BBC News, Mar. 17, 2021, https://www.bbc.co.uk/news/world -asia-56425115.

260   Lori Hinnant and Sam Mednick, "Virus-Linked Hunger Tied to 10,000 Child Deaths Each Month," AP, Jul. 27, 2020, https://apnews.com/article /virus-outbreak-africa-ap-top-news-understanding-the-outbreak-hunger-5c ee9693c52728a3808f4e7b4965cbd.

261   Global Financing Facility, "Emerging Data Estimates that for Each COVID-19 Death, More than Two Women and Children Have Lost Their Lives as a Result of Disruptions to Health Systems Since the Start of the Pandemic," press release, Sep. 29, 2021, https://www.globalfinancingfacility .org/emerging-data-estimates-each-covid-19-death-more-two-women-and -children-have-lost-their-lives-result.

262   "Recessions and Mortality: a Global Perspective," BIS Working Papers, Dec. 15, 2020, https://www.bis.org/publ/work910.htm.

263   Holf, "Haiti Did Not Vaccinate Its Citizens."

264   Tiffany N. Ford, Sarah Reber, & Richard V. Reeves, "Race Gaps in Covid-19 Deaths Are Even Bigger Than They Appear," The Brookings Institution, Jun. 16, 2020, https://www.brookings.edu/blog/up-front /2020/06/16/race-gaps-in-covid-19-deaths-are-even-bigger-than-they -appear.

265   Steven H. Woolf, Ryan K. Masters, Laudan Y. Aron, "Effect of the COVID-19 Pandemic in 2020 on Life Expectancy Across Populations in the USA and Other High Income Countries: Simulations of Provisional

Mortality Data," The BMJ 373, no. 373 (2021): 1343, https://doi.org
/10.1136/bmj.n1343.

266   "COVID-19 and Child Labour: A Time of Crisis, A Time to Act," UNICEF,
       2020, https://www.unicef.org/media/70261/file/COVID-19-and-Child
       -labour-2020.pdf

267   Mary Kekatos, "Maternal Mortality Rates Increased During 1st Year of
       COVID Pandemic: CDC," *ABC News,* Feb. 23, 2022, https://abcnews.
       go.com/Health/maternal-mortality-rates-increased-1st-year-covid-pandemic
       /story?id=83061990.

268   Donna Hoyert, "Maternal Mortality Rates in the United States, 2020,"
       NCHS Health E-Stats, Feb. 2022, https://dx.doi.org/10.15620/cdc:113967.

269   Kathleen H. Krause, Jorge V. Verlenden, Leigh E. Szucs, et al., "Disruptions
       to School and Home Life among High School Students During the
       COVID-19 Pandemic — Adolescent Behaviors and Experiences Survey,
       United States, January–June 2021," *MMWR* Suppl 71, no. 3 (2022): 28–34,
       http://dx.doi.org/10.15585/mmwr.su7103a5.

270   "Children, Obesity, and COVID-19," CDC, last rev. Feb. 18, 2022, https:
       //www.cdc.gov/obesity/data/children-obesity-COVID-19.html.

271   Jiaxing Wang, MD, PhD, Ying Li, MD, PhD, David C. Musch, PhD,
       MPH, et al., "Progression of Myopia in School-Aged Children after
       COVID-19 Home Confinement," *JAMA Ophthalmol.* 139, no. 3, (2021):
       293-300, doi.org/10.1001/jamaophthalmol.2020.6239.

272   Yuki, Noguchi, "Obesity Rates Rise During Pandemic, Fueled by Stress, Job
       Loss, Sedentary Lifestyle," *NPR,* Sep. 29, 2021, https://www.npr.org/sections
       /health-shots/2021/09/29/1041515129/obesity-rates-rise-during-pandemic
       -fueled-by-stress-job-loss-sedentary-lifestyle.

273   "Transmission of SARS-CoV-2: Implications for Infection Prevention
       Precautions," *World Health Organization,* Jul. 9, 2020, https://www.who.int
       /news-room/commentaries/detail/transmission-of-sars-cov-2-implications
       -for-infection-prevention-precautions.

274   Ellen Barry, "Many Teens Report Emotional and Physical Abuse by Parents
       during Lockdown," *New York Times,* Mar. 31, 2022, https://www.nytimes.com
       /2022/03/31/health/covid-mental-health-teens.html.

275   Hoover Institute, "What Happened: Dr. Jay Bhattacharya on 19 Months
       of COVID," *Uncommon Knowledge with Peter Robinson,* Oct. 21, 2021,
       https://www.hoover.org/research/what-happened-dr-jay-bhattacharya-19
       -months-covid-1. *See embedded video at 00:34:50.

276   James Freeman, "The Limits of Anthony Fauci's Expertise," *Wall Street
       Journal,* May 13, 2020, https://www.wsj.com/articles/the-limits-of-anthony
       -faucis-expertise-11589392347.

277   CNBC Television, "Dr. Anthony Fauci and Sen. Rand Paul Debate Reopening
       Schools," YouTube, 00:02:41, May 12, 2020, https://www.youtube.com
       /watch?v=9oxp3wCCxX0.

278   The World Bank, UNESCO, and UNICEF, "The State of the Global
       Education Crisis: A Path to Recovery The State of the Global Education
       Crisis: A Path to Recovery," UNESCO, (2021), https://unesdoc.unesco.org
       /ark:/48223/pf0000380128.

279   Morbidity and Mortality Weekly Report (MMWR), "Adolescent
       Behaviors and Experiences Survey—United States, January–June 2021,"
       CDC Supplement 71, No. 3, Apr. 1, 2022, https://www.cdc.gov/mmwr/
       volumes/71/pdfs/su7103a1-a5-H.pdf.

280   Jessica Grose, "Teenagers Report Growing Anxiety. Maybe That's Rational,"
       New York Times, Apr. 2, 2022, https://www.nytimes.com/2022/04/02/opinion
       /cdc-teenagers.html.

281   Liz Tung, "'We Are in a State of Emergency': What's Behind the Rising
       Suicide Rate among Black Kids," *WHYY PBS-NPR*, Aug. 20, 2021, https://
       whyy.org/segments/we-are-in-a-state-of-emergency-whats-behind-the-rising
       -suicide-rate-among-black-kids.

282   National Center for Health Statistics, "The Record Increase in Homicide
       during 2020," CDC, Oct. 8, 2021, https://www.cdc.gov/nchs/pressroom
       /podcasts/2021/20211008/20211008.htm.

283   Olga Khazan, "Why People Are Acting So Weird," *The Atlantic*, Mar. 30,
       2022, https://www.theatlantic.com/politics/archive/2022/03/antisocial
       -behavior-crime-violence-increase-pandemic/627076.

284   Carly Thomas, "How Masks Could Affect Speech and Language
       Development in Children," *CBC News*, Mar. 17, 2021, https://www.cbc.ca
       /news/science/children-masks-language-speech-faces-1.5948037.

285   Branwen Jeffries, "Lockdowns Hurt Child Speech and Language Skills -
       Report," *BBC*, Apr. 27, 2021, https://www.bbc.com/news/education
       -56889035.

286   Sarah D. Sparks, "More Than 1 in 3 Children Who Started School in the
       Pandemic Need 'Intensive' Reading Help," *Education Week*, Feb. 16, 2022,
       https://www.edweek.org/teaching-learning/more-than-1-in-3-children-who
       -started-school-in-the-pandemic-need-intensive-reading-help/2022/02.

287   Natalie Wexler, "Covid-Era Babies Are 'Talking' Less, Signaling Future
       Reading Challenges," *Forbes*, May 10, 2022, https://www.forbes.com/sites
       /nataliewexler/2022/05/10/covid-era-babies-are-talking-less-signaling-future
       -reading-challenges/?sh=7fa8fd09280a.

288   Ibid.

289   Sean CL Deoni, Jennifer Beauchemin, Alexandra Volpe, Viren D'Sa, and
       the RESONANCE Consortium, "Impact of the COVID-19 Pandemic

on Early Child Cognitive Development: Initial Findings in a Longitudinal Observational Study of Child Health," *medRxiv*, Aug. 11, 2021, https://www.medrxiv.org/content/10.1101/2021.08.10.21261846v1.full.pdf.

290   Ibid.

291   Lauren C. Shuffrey, Morgan R. Firestein, Margaret Kyle, et al, "Association of Birth During the COVID-19 Pandemic with Neurodevelopmental Status at 6 Months in Infants with and without In Utero Exposure to Maternal SARS-COV-2 Infection," *JAMA Pediatrics*, Jan. 4, 2022, https://jamanetwork.com/journals/jamapediatrics/fullarticle/2787479.

292   Camille Chambonniere, Nicole Fearnbach, Lena Pelissier, et al., "Adverse Collateral Effects of COVID-19 Public Health Restrictions on Physical Fitness and Cognitive Performance in Primary School Children," *International Journal of Environmental Research and Public Health*, Sep. 30, 2021, https://www.mdpi.com/1660-4601/18/21/11099/htm.

293   "CDC's Developmental Milestones Checklist," CDC (updated: Feb. 2022), https://www.cdc.gov/ncbddd/actearly/pdf/FULL-LIST-CDC_LTSAE-Checklists2021_Eng_FNL2_508.pdf.

294   Steve Schering, "CDC, AAP Update Developmental Milestones for Surveillance Program," *American Academy of Pediatrics (AAP) News*, Feb. 8, 2022, https://publications.aap.org/aapnews/news/19554/CDC-AAP-update-developmental-milestones-for.

295   David Cutler, PhD, and Lawrence Summers, PhD, "The COVID-19 Pandemic and the $16 Trillion Virus," *JAMA* 324, no. 15 (2020): 1495–1496, www.doi.org/10.1001/jama.2020.19759.

296   Kevin Orland and Brian Platt, "Banks Get Protesters' Names as Canada Financial Squeeze Unfolds," *Bloomberg*, Feb. 17, 2022, https://www.bloomberg.com/news/articles/2022-02-17/police-begin-financial-squeeze-trudeau-defends-emergency-edict.

297   Joanna Ossinger and Carolynn Look, "Crypto Revolution Spurs Central Banks to Design Money's Future," *Bloomberg*, Apr. 14, 2022, https://www.bloomberg.com/news/features/2022-04-14/crypto-in-europe-china-nigeria-turn-to-cbdc-digital-cash-experiments.

298   U.S. Congressional Record – Senate, "89th Congress, 2nd Session Vol. 112, Part 8 – Bound Edition," *United States Congress*, May 9, 1966, p. 40, https://www.congress.gov/89/crecb/1966/05/09/GPO-CRECB-1966-pt8-4-2.pdf.

299   Alexander Rubenstein and Max Blumenthal, "How Ukraine's Jewish President Zelensky Made Peace with Neo-Nazi Paramilitaries on Front Lines of War with Russia," *The Grayzone*, Mar. 4, 2022, https://thegrayzone.com/2022/03/04/nazis-ukrainian-war-russia.

300   Khaleda Rahman, "Joe Biden Mentions Oil Only Once in Essay Defending
       Saudi Arabia Trip," *Newsweek*, Jul. 10, 2022, https://www.newsweek.com
       /joe-biden-mentions-oil-only-once-essay-defending-saudi-arabia-trip
       -1723204.

301   "Yemen War Deaths Will Reach 377,000 by End of the Year: UN," *Al
       Jazeera*, Nov. 23, 2021, https://www.aljazeera.com/news/2021/11/23/un
       -yemen-recovery-possible-in-one-generation-if-war-stops-now.

302   William Roberts, "MBS Approved Operation to Capture or Kill Khashoggi:
       US Report," *Al Jazeera*, Feb. 26, 2021, https://www.aljazeera.com/
       news/2021/2/26/mbs-oversaw-saudi-killers-of-khashoggi-us-intel-report.

303   Marty Makary, "The CDC — Which Is Withholding Information—Has a
       Hidden Agenda," *New York Post*, Feb. 27, 2022, https://nypost.com/2022
       /02/27/the-cdc-has-a-hidden-agenda-when-it-comes-to-covid-vaccines.